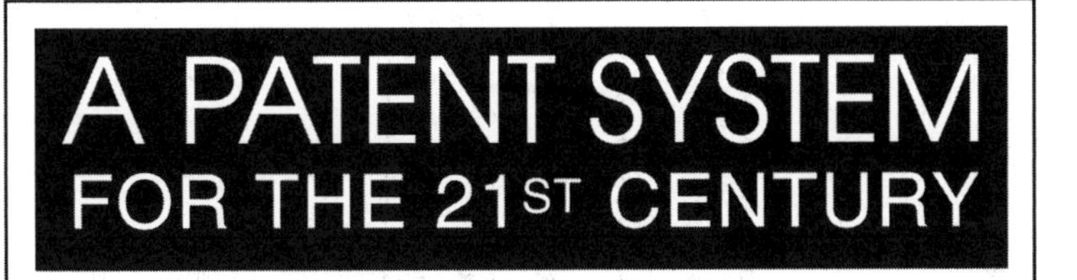

A PATENT SYSTEM FOR THE 21ST CENTURY

Stephen A. Merrill, Richard C. Levin, and Mark B. Myers, Editors

Committee on Intellectual Property Rights in the Knowledge-Based Economy
Board on Science, Technology, and Economic Policy
Policy and Global Affairs Division

NATIONAL RESEARCH COUNCIL
OF THE NATIONAL ACADEMIES

THE NATIONAL ACADEMIES PRESS
Washington, D.C.
www.nap.edu

THE NATIONAL ACADEMIES PRESS • 500 Fifth Street, N.W. • Washington, DC 20001

This study was supported by Contract No. NASW-99037, Task Order 103, between the National Academy of Sciences and the National Aeronautics and Space Administration, and by the U.S. Department of Commerce, Andrew W. Mellon Foundation, Center for the Public Domain, Pharmacia Corporation, Merck & Company, Procter & Gamble, and IBM. Any opinions, findings, conclusions, or recommendations expressed in this publication are those of the author(s) and do not necessarily reflect the views of the organizations or agencies that provided support for the project.

International Standard Book Number 0-309-08910-7 (Book)
International Standard Book Number 0-309-50610-7 (PDF)
Library of Congress Control Number TK

Limited copies are available from:

Board on Science, Technology, and Economic Policy
National Research Council
500 Fifth Street, N.W.
Washington, DC 20001
Phone: 202-334-2200
Fax: 202-334-1505

Additional copies of this report are available from National Academies Press, 500 Fifth Street, N.W., Lockbox 285, Washington, DC 20055; (800) 624-6242 or (202) 334-3313 (in the Washington metropolitan area); Internet, http://www.nap.edu.

Printed in the United States of America

Companion volume:
Patents in the Knowledge-Based Economy, edited by Wesley M. Cohen and Stephen A. Merrill, studies patent quality, litigation, and patenting and licensing in biotechnology, software, and the business methods.

The cover design incorporates original illustration from the following U.S. patents issued over a nearly 160-year period:

U.S. Patent 6,506,554; Core structure of gp41 from the HIV envelope glycoprotein; Chan, David C. (Brookline, MA); Fass, Deborah (Cambridge, MA); Lu, Min (New York, NY); Berger, James M., (Cambridge, MA); Kim, Peter S. (Lexington, MA); Granted January 14, 2003.
U.S. Patent 6,423,583; Methodology for electrically induced selective breakdown of nanotubes; Avouris, Phaedon (Yorktown Heights, NY); Collins, Philip G. (Ossining, NY); Martel, Richard (Peekskill, NY); Granted July 23, 2003.
U.S. Patent 6,313,562; Microelectromechanical ratcheting apparatus; Barnes, Stephen M. (Albuquerque, NM); Miller, Samuel L. (Albuquerque, NM); Jensen, Brian D. (Albuquerque, NM); Rodgers, M. Steven (Albuquerque, NM); Burg, Michael S., (Albuquerque, NM); Granted November 6, 2001.
U.S. Patent 821,393; Flying machine; Wright , Orville (Dayton, OH) and Wright, Wilbur (Dayton, OH); Granted May 22, 1906.
U.S. Patent 223,898; Electric lamp; Edison, Thomas A. (Menlo Park, NJ); Granted January 27, 1880.
U.S. Patent 4750; Improvement in sewing machines; Howe, Jr., Elias, (Cambridge, MA); Granted September 10, 1846.

THE NATIONAL ACADEMIES

Advisers to the Nation on Science, Engineering, and Medicine

The **National Academy of Sciences** is a private, nonprofit, self-perpetuating society of distinguished scholars engaged in scientific and engineering research, dedicated to the furtherance of science and technology and to their use for the general welfare. Upon the authority of the charter granted to it by the Congress in 1863, the Academy has a mandate that requires it to advise the federal government on scientific and technical matters. Dr. Bruce M. Alberts is president of the National Academy of Sciences.

The **National Academy of Engineering** was established in 1964, under the charter of the National Academy of Sciences, as a parallel organization of outstanding engineers. It is autonomous in its administration and in the selection of its members, sharing with the National Academy of Sciences the responsibility for advising the federal government. The National Academy of Engineering also sponsors engineering programs aimed at meeting national needs, encourages education and research, and recognizes the superior achievements of engineers. Dr. Wm. A. Wulf is president of the National Academy of Engineering.

The **Institute of Medicine** was established in 1970 by the National Academy of Sciences to secure the services of eminent members of appropriate professions in the examination of policy matters pertaining to the health of the public. The Institute acts under the responsibility given to the National Academy of Sciences by its congressional charter to be an adviser to the federal government and, upon its own initiative, to identify issues of medical care, research, and education. Dr. Harvey V. Fineberg is president of the Institute of Medicine.

The **National Research Council** was organized by the National Academy of Sciences in 1916 to associate the broad community of science and technology with the Academy's purposes of furthering knowledge and advising the federal government. Functioning in accordance with general policies determined by the Academy, the Council has become the principal operating agency of both the National Academy of Sciences and the National Academy of Engineering in providing services to the government, the public, and the scientific and engineering communities. The Council is administered jointly by both Academies and the Institute of Medicine. Dr. Bruce M. Alberts and Dr. Wm. A. Wulf are chair and vice chair, respectively, of the National Research Council.

www.national-academies.org

Committee on Intellectual Property Rights in the Knowledge-Based Economy

Board on Science, Technology, and Economic Policy

Richard C. Levin, Co-chair
President
Yale University

Mark B. Myers, Co-chair
Visiting Executive Professor of Management
The Wharton School
University of Pennsylvania

John H. Barton
George E. Osborne Professor of Law Emeritus
Stanford University Law School

Robert Blackburn
Vice President and Chief Patent Counsel
Chiron Corporation
and
Distinguished Scholar
Berkeley Center for Law and Technology

Wesley Cohen
Professor of Economics and Management
Fuqua School of Business
Duke University

Frank Collins
Senior Vice President for Research
ZymoGenetics

Rochelle Cooper Dreyfuss
Pauline Newman Professor of Law
New York University School of Law

Bronwyn H. Hall
Professor of Economics
University of California, Berkeley

Eugene Lynch
Judge (ret.)
U.S. District Court for the Northern District of California

Daniel P. McCurdy
President & CEO
ThinkFire, Ltd.

Gerald J. Mossinghoff
Senior Counsel
Oblon, Spivak, McClelland, Maier & Neustadt, P.C.
(until December 2003)

Gail K. Naughton
Dean, School of Business
San Diego State University

Richard R. Nelson
George Blumenthal Professor of International and Public Affairs
Columbia University

James Pooley
Partner
Milbank, Tweed, Hadley & McCloy LLP

William J. Raduchel
Great Falls, Virginia

Pamela Samuelson
Chancellor's Professor of Law and Information Management
and
Director, Berkeley Center for Law and Technology
University of California, Berkeley

Liaisons with Other NRC Programs

Science, Technology and Law Program
David Korn
Senior Vice President for Biomedical and Health Sciences Research
Association of American Medical Colleges

National Cancer Policy Board

Pilar Ossorio
Assistant Professor of Law and Medical Affairs
University of Wisconsin Law School

Staff

Stephen A. Merrill
Project Director

Craig Schultz
Research Associate

Camille Collett
Project Associate (until September 2002)

George Elliott
U.S. Department of Commerce Science and Technology Fellow (from September 2000 through September 2001)

Russell Moy
Senior Program Officer (from June 2002)

Aaron Levine
NRC Intern (Summer 2003)

Peter Kozel
NRC Intern (Winter-Spring 2004)

Staff

Stephen A. Merrill
Executive Director

Charles Wessner
Deputy Director

Russell Moy
Senior Program Officer

Sujai Shivakumar
Program Officer

Craig Schultz
Research Associate

McAlister Clabaugh
Program Associate

David Dierksheide
Program Associate

Amy Przybocki
Financial Associate

Acknowledgments

Our committee's study of the patent system was a much more ambitious undertaking than we anticipated at its outset, and we have many people to thank for their contributions to its completion. First, through eight meetings, two conferences, numerous report drafts, and preparation of the response to reviewers' comments, the members of the committee not only provided thoughtful individual contributions but also successfully bridged differences in professional training and experience to reach a common understanding and consensus recommendations. One committee member, Gerald Mossinghoff, resigned on December 1, 2003, as this report was being revised before submission to the National Research Council review process. He played a very active and constructive role in the deliberations of the committee and provided comments on a preliminary report draft. We regret not having the benefit of his advice in the final stage of writing.

More than 150 people assisted the committee's deliberations in a variety of ways—conducting and reporting on research, speaking at conferences, presenting views at open meetings of the committee, and providing other valuable information through communications with staff. Their contributions were indispensable to the committee's work, and they are listed in Appendix B of the report.

Although self-initiated, the study as a whole or activities within it have attracted diverse support from government agencies, foundations, and corporations. The National Aeronautics and Space Administration sponsored the project as part of its program support of the Board on Science, Technology, and Economic Policy (STEP) from 1999 to 2003. The Andrew W. Mellon Foundation was principal sponsor of the conference *Intellectual Property Rights: How Far Should They Be Extended?* as well as of the commissioned research activities that followed it. A supplemental Mellon contribution will help to support dissemination of the results of the project. A grant from the Center for the Public Domain enabled the project to develop a web site dedicated to intellectual property issues across the National Academies and supported other activities. The web site (http://ip.nationalacademies.org) has been an indispensable part of our efforts to keep the community of interested people informed of our progress and an avenue for them to express their views. The U.S. Department of Commerce, through its Technology Administration, sponsored a conference on university patenting and licensing. Several corporations—Pharmacia, Merck and Co., Procter and Gamble, and IBM—provided unrestricted funds. Finally, the U.S. Patent and Trademark Office (USPTO) paid the salary of a senior patent examining supervisor who, under the Commerce Department's Science and Technology Fellowship Program, worked full-time with the staff and committee from September 2000 through September 2001. The USPTO also provided data and suggested factual corrections to the prepublication version of this report at our request. We are very grateful to these sponsors and contributors.

At the outset of the study phase of this project, we were encouraged to consult other National Research Council boards and committees with an interest in intellectual property policy and relevant technical and legal expertise. The committee extended invitations to three program units to select a volunteer to serve in a liaison capacity, contributing to framing the study and participating in its fact-finding phrase but not deliberating on nor assuming responsibility for the committee's findings and recommendations. Two units accepted STEP's invitation. David Korn of the Association of American Medical Colleges represented the Science, Technology, and Law Program and Pilar Ossorio of the University of Wisconsin law faculty represented the National Cancer Policy Board of the Institute of Medicine. Both actively participated in a number of panel meetings, helped to focus the inquiry in its early stages, and provided useful information to the committee. We are grateful to them for their assistance, especially as they were not able to see the project through to its conclusion.

This report has been reviewed in draft form by individuals chosen for their diverse perspectives and technical expertise, in accordance with procedures approved by the National Research Council's Report Review Committee. The purpose of this independent review is to provide candid and critical comments that will assist the institution in making its published report as sound as possible and to ensure that the report meets institutional standards for objectivity, evidence, and responsiveness to the study charge. The review comments and draft manuscript remain confidential to protect the integrity of the deliberative process. We wish to thank the following individuals for their review of this report:

Robert Armitage, Eli Lilly & Co.
David Beier, Amgen, Inc.
Q. Todd Dickinson, Howrey, Simon, Arnold & White
Raoul Drapeau, Products Concepts Co.

David Forney, Motorola, Inc. (ret.)
Dominique Guellec, European Patent Office
Donald Grant Kelley, Intellectual Asset Management Assoc., LLC
Katharine Ku, Stanford University
Roderick McKelvie, Fish & Neave
Robert Merges, University of California, Davis, Law School
Steven Odre, Amgen, Inc.
Arti Rai, Duke University Law School
F. M. Scherer, Harvard University emeritus
Carl Shapiro, University of California, Berkeley
Anthony Siegman, Stanford University emeritus

Although the reviewers provided many constructive comments and suggestions, they were not asked to endorse the conclusions or recommendations nor did they see the final draft of the report before its release. The review of this report was overseen by Gilbert Omenn, University of Michigan, and Joe Cecil, Federal Judicial Center. Appointed by the National Research Council, they were responsible for ensuring that an independent examination of this report was carried out in accordance with institutional procedures and that all review comments were carefully considered. Responsibility for the final content of this report rests entirely with the authoring committee and the institution.

Our most profound thanks are reserved for STEP's executive director, Stephen Merrill, and his excellent staff. Steve made a prodigious personal effort to assist the committee even as he continued to manage the office and supervise a number of other projects. We are grateful that he took responsibility for writing the first drafts of everything and impressed by how patiently he integrated the comments, suggestions, and substitute prose of committee members and outside reviewers. His good sense and evenhandedness prevailed throughout our committee's lively and often contentious internal debates, which persisted until the final draft was sent to the printer.

Craig Schultz provided superb research, production, and administrative support throughout the project. He created the National Academies' intellectual property web site and produced on CD-ROM the conference proceedings as well as the research collection and the final report. The success of our conferences, workshops, and public hearings is a tribute to his organizational and personal skills.

During the crucial first year of our deliberations the committee had the benefit of the experience, detailed knowledge, and good sense of George Elliott, on leave from the USPTO as a Commerce science and technology fellow with the National Academies. To the extent that our report reflects some understanding of the details of the patent examination process and life of examiners, George is responsible. He is responsible neither for our errors of fact nor the flaws in our recommendations.

Executive Summary

Since its creation more than 200 years ago, the U.S. patent system has played an important role in stimulating technological innovation by providing legal protection to inventions of every description and by disseminating useful technical information about them. With the growing importance of technology to the nation's well-being, patents are playing an even more prominent role in the economy. There are many indications that firms of all sizes as well as universities and public institutions are ascribing greater value to patents and are willing to pay higher costs to acquire, exercise, and defend them.

Throughout its history the patent system has had to adapt to evolving conditions, and it continues to demonstrate flexibility and responsiveness today. Since 1980 a series of judicial, legislative, administrative, and diplomatic actions have extended patenting to new technology (biotechnology) and to technologies previously without or subject to other forms of intellectual property protection (software), encouraged the emergence of new players (universities and public research institutions), strengthened the position of patent holders vis-à-vis alleged infringers domestically and internationally, relaxed antitrust constraints on the use of patents, and extended the reach of patenting upstream from commercial products to scientific research tools, materials, and discoveries.

Continuing high rates of innovation suggest that the patent system is working well and does not require fundamental changes. We generally agree with that conclusion, but it is clear that both economic and legal changes are putting new strains on the system. Patents are being more actively sought and vigorously enforced. The sheer volume of applications to the U.S. Patent and Trademark Office—more than 300,000 a year—threatens to overwhelm the patent examination corps, degrading the quality of their work or creating a huge backlog of pending cases, or both. The costs of acquiring patents, promoting or securing licenses to patented technology, and defending against infringement allegations in court are rising rapidly. The benefits of patents in stimulating innovation appear to be highly variable across technologies and industries, but there has been little systematic investigation of the differences. In some cases patenting appears to have departed from its traditional role, as firms build large portfolios to gain access to others' technologies and reduce their vulnerability to litigation.

In light of these strains, now is an opportune time to examine the system's performance and consider how it can continue to reinvent itself. In spite of its pervasive influence, patent policy for the last 50 years has been the preserve of practicing attorneys, judges, patent office administrators, and legally trained legislators. The National Academies believe that patent policy will benefit from the additional insights of economists, scientists, and engineers in different disciplines, inventors, business managers, and legal scholars, and they appointed our committee to reflect that diversity of expertise.

We in turn benefited from the insights and data of nine groups of scholars supported by the National Research Council's Board on Science, Technology, and Economic Policy (STEP) to conduct a series of policy-related empirical studies. These are collected in this report's companion volume, *Patents in the Knowledge-Based Economy*. This work is part of a growing body of economic and legal research since 1980. Still, it is quite limited, and the range of industries examined in any detail is quite narrow. We do not know whether the benefits of more and "stronger" patents extend very far beyond a few manufacturing industries, such as pharmaceuticals, chemicals, and medical devices. It is even less clear that patents induce additional research and development investment in the service industries and service functions of the manufacturing economy. One obvious conclusion of our work is that we need a much more detailed understanding of how the patent system affects innovation in various sectors. But even without additional study we can identify areas of strain, inefficiency, excessive cost on the one hand and inadequate resources on the other hand that need to be addressed now.

CRITERIA FOR EVALUATING THE PATENT SYSTEM

In circumstances that at this stage defy a comprehensive evaluation of the patent system's impact on innovation, we identify seven performance criteria that are widely thought to be important if not necessary conditions for innovation and that are in some degree measurable.

First Criterion: ***The patent system should accommodate new technologies.*** The U.S. patent system has excelled at adapting to change because it is a unitary system with few *a priori* exclusions. The initiative to extend patenting to new areas lies in the first instance with inventors and commercial developers rather than legal authorities, and the system, while formally neutral, has features that allow for somewhat different treatment of different technologies.

The incorporation of emerging technologies is not always seamless and rapid; indeed, it often generates considerable controversy. Moreover, case law recognizes limits to patenting, confining patents to inventions that can be expressed as products or methods and excluding patents on abstract ideas and phenomena of nature. Some, although not all, members of the committee are concerned that recent fairly abstract patents cross this indistinct line and have unwisely limited public access to ideas and techniques that are important to basic scientific research.

Second Criterion: ***The system should reward only those inventions that meet the statutory tests of novelty and utility, that would not at the time they were made be obvious to people skilled in the respective technologies, and that are adequately described.*** Over the past decade the quality of issued patents has come under frequent sharp attack, as it sometimes has in the past. Some critics have suggested that the standards of patentability—especially the non-obviousness standard—have become too lax as a result of court decisions. Other observers fault the performance of the U.S. Patent and Trademark Office (USPTO) in examining patent applications, variously attributing the alleged deterioration to inadequate time for examiners to do their work, lack of access to prior art information, or the qualifications of the corps of examiners.

The claim that quality has deteriorated in a broad and systematic way could be, but has not been, empirically tested. Therefore, conclusions must remain tentative. There are nevertheless several reasons to suspect that more issued patents are substandard, particularly in technologies newly subject to patenting. One reason to believe that quality has suffered, even before taking examiner qualifications and experience into account, is that in recent years the number of patent examiners has not kept pace with the increase in workload represented by the escalating number and growing complexity of applications. Second, according to recent estimates taking into account patent continuations, overall patent approval rates appear to be higher than officially reported, and at least in the past few years have been higher than in the European and Japanese patent offices. Third, changes in the treatment of genomic and business method applications, introduced as a result of criticisms of the quality of patents being issued, has reduced or at least slowed down the number of patent grants in those fields. And fourth, there might have been some dilution of the application of the non-obviousness standard in biotechnology and some limitations on its proper application to business methods patent applications. Although quality appears to be more problematic in rapidly moving areas of technology newly subject to patenting and is perhaps corrected over time, the cost of waiting for an evolutionary process to run its course may be too high when new technologies attract the level of investment exhibited by the Internet and biotechnology.

Third Criterion: ***The patent system should serve its second function of disseminating technical information.*** In the United States there are many channels of scientific interaction and technical communication, and the patent system contributes more than does the alternative of maintaining technical advances as trade secrets. There are, nonetheless, features peculiar to the U.S. patent system that inhibit information dissemination. One is the exclusion of about 10 percent of U.S. patent applications from publication, although universal publication 18 months after filing has been an international norm since 1994. A second U.S. idiosyncrasy is the legal doctrine of willful infringement, which can require an infringer to pay triple damages if it can be demonstrated that the infringer was aware of the patent before the infringement. Some observers believe that this deters an inventor from looking at the patents of possible competitors, because knowledge of the patent could later make the inventor subject to triple damages if there were an infringement case. This undermines one of the principal purposes of the patent system—to make others aware of innovations that could help stimulate further innovation.

Fourth Criterion: ***Administrative and judicial decisions entailed in the patent system should be timely, and the costs associated with them should be reasonable and proportionate.*** The elapsed time between the filing of a patent application and the patent examiner's first action on it and the time between filing and final disposition are lengthening, particularly in new technologies, although resolution takes longer in other countries than in the United States. By the same token, it takes an inordinately long time to resolve questions of patent validity in the courts, and the cost of the proceeding is escalating. The burden of costs and uncertainties, especially those entailed in challenging and defending patents, falls disproportionately on smaller, less experienced firms.

Fifth Criterion: ***Access to patented technologies is important in research and in the development of cumulative technologies, where one advance builds upon one or several previous advances.*** Faced with anecdotes and conjectures about restrictions on researchers, particularly in biotechnology, the committee initiated a modest, interview-based survey of diverse participants in the field to determine whether patent thickets were emerging or access to foundational discoveries was restricted. The results suggest that intellectual property in biotechnology is being managed relatively successfully. The associated costs are somewhat higher and research can sometimes be slowed, but it is rarely blocked altogether. There are, however, occasional cases of restricted access to foundational discoveries and to some diagnostic genetic tests. Universities have traditionally operated under an unwritten assumption that they would not be sued by patent holders for violating patents in the course of precommercial university research, but a ruling in 2002 by the U.S. Court of Appeals for the Federal Circuit made it clear that a university is not legally protected from patent infringement liability. It remains to be seen whether this will change the behavior of patent holders toward university research, but universities are at greater risk.

Sixth Criterion: ***Greater integration of or reciprocity among the three major patent systems would reduce public and private transaction costs, facilitating trade, investment, and innovation.*** In spite of progress in harmonizing the U.S., European, and Japanese patent examination systems, important differences in standards and procedures remain, ensuring search and examination redundancy that imposes high costs on users and hampers market integration. These include differences with respect to assigning patent application priority, the requirement to disclose a technology's best implementation to qualify for a patent, the period, if any, allowed between publication of an invention and submission of a patent application, and whether all patent applications are published after 18 months.

Seventh Criterion: ***There should be a level field, with intellectual property rights holders who are similarly situated (e.g., state and private institutions performing research) enjoying the same benefits while being subject to the same obligations.*** In 1999 the Supreme Court struck down a law that denied a state's ability under the Eleventh Amendment to the Constitution to claim immunity against charges of infringing a patent or other intellectual property. Under the ruling a state institution such as a public university holding a patent could be in the position of asserting its patent rights against an infringer while successfully barring a patent holder from recovering damages for the university's infringement of a patent although the state institution might be enjoined from further infringement. A private university enjoys no protection from infringement suits. Although it is too soon to know what the effects of the Supreme Court decision will be, one possibility is that the disparity could influence decisions on where research is done.

RECOMMENDATIONS TO IMPROVE THE PATENT SYSTEM

The committee supports seven steps to ensure the vitality and improve the functioning of the patent system:

1. Preserve an open-ended, unitary, flexible patent system. The system should remain open to new technologies, and the features that allow somewhat different treatment of different technologies should be preserved without formalizing different standards, for example, in statutes that would be exceedingly difficult to draft appropriately, difficult to change if found to be antiquated or inappropriate, and at odds with U.S. international commitments. Among the tailoring mechanisms that should be fully exploited is the USPTO's development of examination guidelines for new or newly patented technologies, as has been done for computer programs, superconductivity, and genetic inventions. In developing such guidelines the office should seek advice from a wide variety of sources and maintain a public record of the submissions, and the results should be part of the record of any appeal to a court so that they can inform judicial decisions.

This information could be of particular value to the Court of Appeals for the Federal Circuit, which is in most instances the final arbiter of patent law. Further, in order for the judges to keep themselves well informed about relevant legal and economic scholarship, the court should encourage the submission of amicus briefs and arrange for temporary exchanges of members with other courts. Appointments to the Federal Circuit should include people familiar with innovation from a variety of perspectives, including management, finance, and economic history, as well as nonpatent areas of law that could have an effect on innovation.

2. Reinvigorate the non-obviousness standard. The requirement that to qualify for a patent an invention cannot be obvious to a person of ordinary skill in the art should be assiduously observed. In an area such as business methods, where the common general knowledge of practitioners is not fully described in published literature likely to be consulted by patent examiners, another method of determining the state of knowledge needs to be employed. Given that patent applications are examined *ex parte* between the applicant and the examiner, it would be difficult to bring in other expert opinions at that stage. Nevertheless, the Open Review procedure described below provides a means of obtaining expert participation if a patent is challenged.

Gene sequence patents present a particular problem because of a Federal Circuit ruling whose practical effect was to make it difficult to make a case of obviousness against a biological macromolecule claimed by its structure. This is unwise in its own right and is also inconsistent with patent practice in other countries. The court should return to a standard that would not grant a patent for an innovation that any skilled colleague would also have tried with a "reasonable expectation of success."

3. Institute an Open Review procedure. Congress should seriously consider legislation creating a procedure for third parties to challenge patents after their issuance in a proceeding before administrative patent judges of the USPTO. The grounds for a challenge could be any of the statutory standards—novelty, utility, non-obviousness, disclosure, or enablement—or even the case law proscription on patenting abstract ideas and natural phenomena. The time, cost, and other characteristics of this proceeding should make it an attractive alternative to litigation to resolve patent validity questions both for private disputants and for federal district courts. The courts could more productively focus their attention on patent infringement issues if they were able to refer validity questions to an Open Review proceeding.

4. Strengthen USPTO capabilities. To improve its performance the USPTO needs additional resources to hire and train additional examiners and fully implement a robust electronic processing capability. Further, the USPTO should create a strong multidisciplinary analytical capability to assess management practices and proposed changes, provide an early warning of new technologies being proposed for patenting, and conduct reliable, consistent, reputable quality reviews that address office-wide and individual examiner performance. The current USPTO budget is not adequate to accomplish these objectives, let alone to finance an efficient Open Review system.

5. Shield some research uses of patented inventions from liability for infringement. In light of the Federal Circuit's 2002 ruling that even noncommercial scientific research conducted in a university enjoys no protection from patent infringement liability and in view of the degree to which the academic research community especially has proceeded with their work in the belief that such an exception existed, there should be limited protection for some research uses of patented inventions. Congress should consider appropriate targeted legislation, but reaching agreement on how this should be done will take time. In the meantime the Office of Management and Budget and the federal government agencies sponsoring research should consider extending "authorization and consent" to those conducting federally supported research. This action would not limit the rights of the patent holder, but it would shift infringement liability to the government. It would have the additional benefit of putting federally sponsored research in state and private universities on the same legal footing without revising the recent Supreme Court's ruling shielding state universities from damage awards in patent infringement suits.

6. Modify or remove the subjective elements of litigation. Among the factors that increase the cost and decrease the predictability of patent infringement litigation are issues unique to U.S. patent jurisprudence that depend on the assessment of a party's state of mind at the time of the alleged infringement or the time of patent application. These include whether someone "willfully" infringed a patent, whether a patent application included the "best mode" for implementing an invention, and whether an inventor or patent attorney engaged in "inequitable conduct" by intentionally failing to disclose all prior art when applying for a patent. Investigating these questions requires time-consuming, expensive, and ultimately subjective pretrial discovery, a principal source of soaring litigation costs. The committee believes that significantly modifying or eliminating these rules would increase the predictability of patent dispute outcomes without substantially affecting the principles that these aspects of the enforcement system were meant to promote.

7. Reduce redundancies and inconsistencies among national patent systems. The United States, Europe, and Japan should further harmonize patent examination procedures and standards to reduce redundancy in search and examination and eventually achieve mutual recognition of results. Differences that need reconciling include application priority ("first-to-invent" versus "first-inventor-to-file"), the grace period for filing an application after publication, the "best mode" requirement of U.S. law, and the U.S. exception to the rule of publication of patent applications after 18 months. This objective should continue to be pursued on a trilateral or even bilateral basis if multilateral negotiations are not progressing.

In making these recommendations the committee is mindful that although the patent law is general, its effects vary across technologies, industries, and classes of inventors. There is a tendency in discourse on the patent system to identify problems and solutions to them from the perspective of one field, sector, or class. Although the committee did not attempt to deal with the specifics of every affected field, the diversity of our membership enabled it to consider each of the proposed changes from the perspective of very different sectors. Similarly, in our deliberations we examined closely the claims made to us that one class of American inventors—individuals and very small businesses—would be disadvantaged by certain changes in the patent system. Some of our recommendations—universal publication of applications, Open Review, and shifting to a first-inventor-to-file system—have in the past been vigorously opposed on those grounds. We conclude that the evidence for such claims is wanting and believe that our recommendations, on balance, would be as beneficial to small entities as to the economy at large.

1

Introduction

Should we revise intellectual property policies and statutes? The best answers will arise when the legal issue is focused by previous conversations among science, business, economics and law. Neither courts nor legislatures may find wise answers in the absence of such interaction.

The Honorable Stephen Breyer, Associate Justice,
Supreme Court of the United States[1]

Over a 10-year period the National Academies' Board on Science, Technology, and Economic Policy (STEP) has investigated a wide range of macro- and microeconomic policies, their impact on investment in research and innovation, and the contribution of research and technology, in turn, to economic performance. In 1999 the board completed a study of the competitive performance of 11 U.S. industrial sectors, in both manufacturing and services. It found that much of the improvement from the 1980s to the 1990s derived from a combination of corporate strategies and public policies supportive of innovation, including steady and conservative fiscal policy, economic deregulation, trade liberalization, lenient antitrust enforcement, and the research investments of previous decades. On the other hand, the board found little evidence, one way or the other, of the economic effects of the many steps taken during the 1980s and 1990s to extend and strengthen intellectual property rights (IPRs). Described in more detail in Chapter 2, these include legislation, court decisions, administrative actions, and international agreements that have resulted in

[1]Speech at the Whitehead Institute, MIT, Cambridge Massachusetts, March 2000, "Genetic Advances and Legal Institutions."

- extending patenting to computer software, genetically modified organisms, nucleic acid molecules, and methods of performing business functions;
- lengthening the duration of copyright protection and extending the term of some patents;
- encouraging universities and nonprofit research institutions to acquire and exercise patent rights;
- strengthening the position of rights holders versus alleged infringers;
- giving federal protection to trade secrets; and
- relaxing antitrust scrutiny of patent use and arrangements.

Curiosity about the effects of these actions led the STEP Board to organize a series of meetings with legal scholars, economists, practitioners, technologists, and corporate managers. In 1999 the board held roundtable discussions in New Haven, Connecticut, hosted by Yale University, and in Berkeley, California, hosted by the Berkeley Center for Law and Technology at Boalt Hall. In February 2000 a two-day STEP conference at the National Academies' headquarters in Washington drew more than 400 participants to discuss *Intellectual Property Rights: How Far Should They Be Extended?*

It was apparent from these discussions that whatever their long-term economic effect, the patent policy changes instituted in the 1980s and 1990s were associated with much more vigorous acquisition, assertion, and enforcement of intellectual property rights than occurred before 1980. Several participants in the meetings, primarily representatives and observers of the information technology and telecommunications sectors, expressed concern about the high costs associated with the acquisition and exercise of IPRs and with the need to develop stronger defensive intellectual property (IP) positions in a litigious environment. Others, primarily academics and representatives of the pharmaceutical industry, voiced concerns about the extension of IP rights to tools and materials of biomedical research, possibly inhibiting the conduct of research and commercial application of its results. A common theme was that the escalation in the number of patents, possibly encouraged by a lowering of the threshold to their acquisition, was creating thickets of rights that could impede innovation by making it difficult or impossible to negotiate access on affordable terms to all of the IP necessary to commercialize a new product or service.

The board concluded that intellectual property policy should be an important part of its agenda and that it should focus initially on the operation of the patent system. The need for specialized legal and technical expertise to carry out a study leading to policy recommendations in these areas led the board to propose to the National Academies the creation of the Committee on Intellectual Property Rights in the Knowledge-Based Economy, composed of economists specializing in intellectual property and technological change, legal scholars, practitioners from corporations and private law practice, biomedical scientists, managers of research and business development in the information technology (IT) sector, a former federal judge, and a former commissioner of the U.S. Patent and Trademark Office (USPTO).[2] The STEP Board and the National Academies' Governing Board charged the committee to

> consider how the resources devoted to patent application review, the standards of patenting, and the patents issued have changed and how these affect incentives to undertake and communicate research and to commercialize new technology. The project will examine how post-patenting patterns of technology licensing and patent infringement litigation affect innovation and diffusion of technology. The study will use evidence from software technology, especially involving Internet business methods, and biotechnology, in particular, genetic sequences. To the extent that current policies and practices serve as a disincentive to research and development and diffusion of new technologies, the study will consider changes in patent administration and dispute resolution processes.

The composition of our study committee is unusual, especially in the recent history of commissions and advisory committees on U.S. patent policy, in the diversity of experience and expertise engaged in evaluating the patent system. Beginning in World War I, the National Research Council (NRC) had a standing Patent Committee composed largely of industrial and academic engineers. Concerned mainly with the role of patents in university and other nonprofit research, it produced one generic report on the functioning of the patent system, issued in 1919 under the acting chairmanship of L. H. Baekeland, inventor and founder of General Bakelite Corporation. A second NRC report, responding to a Department of Commerce request for guidance on how the patent system could more effectively stimulate the growth of new industries, was issued in 1936 by a panel chaired by Vannevar Bush, then vice president and dean of engineering at the Massachusetts Institute of Technology (MIT). The National Patent Planning Commission created by President Franklin D. Roosevelt in 1941 included the inventor of the automobile self-starter, a corporate chief executive officer, a regional Federal Reserve Board official, a labor representative, and a university president. Three more recent publicly appointed panels were narrower in composition, primarily senior managers of Fortune 100 companies, their in-house legal counsel, and members of their outside law firms. These included a presidentially appointed Commission on the Patent System, reporting to Lyndon Johnson in 1966, the 1978 patent policy subcommittee of the Advisory Committee on Industrial

---

[2]Three members of the committee are patent holders, one of them recognized by the Intellectual Property Owners organization as Inventor of the Year 2000. Three members are currently or have recently been involved in the management of entrepreneurial start-up companies, while others have served as directors of such firms. The committee's technical expertise spans a wide spectrum—biotechnology and pharmaceuticals, chemicals, bioengineering, software, microelectronics, and telecommunications.

Innovation at the end of the Carter administration, and the 1992 Advisory Commission on Patent Law Reform, appointed by Commerce Secretary Robert Mosbacher in the first Bush administration. In the mid-1950s the Senate Judiciary Subcommittee on Patents, Trademarks, and Copyrights under Senator Joseph C. O'Mahoney (D-WY) conducted an investigation that drew upon a somewhat more diverse group of experts, including Vannevar Bush and Raymond Vernon, a Harvard international economist. The 2003 report of the Federal Trade Commission (FTC), *To Promote Innovation: The Proper Balance of Competition and Patent Law and Policy,* relied heavily on the testimony of economists and legal scholars as well as practitioners. The principal recommendations of these study panels are listed in Table 1-1.

At the same time that the committee was being assembled, the need for additional analysis and data to inform eventual recommendations led the STEP Board to support nine modest policy-related empirical studies selected from more than 80 proposals submitted in response to a request that was widely circulated in the academic and consulting communities. This research addressed four areas—the functioning of the patent examination process, litigation, and patent acquisition and use in biotechnology and software development. Preliminary results were presented at a Washington, D.C., conference in October 2001, where attorneys, judges, former USPTO officials, and corporate managers commented on the findings. Reviewed and revised papers are included in a companion volume to this report, *Patents in the Knowledge-Based Economy,* edited by Wesley Cohen, a member of the panel, and Stephen Merrill, director of both phases of the project.

The study committee continued the STEP Board's practice of soliciting a wide variety of opinions and airing them in public forums. It met eight times in Washington, D.C.; Woods Hole, Massachusetts; and Palo Alto, California. Apart from the Woods Hole meeting, each event included a public session with testimony from invited speakers and opportunity for observers to comment and raise questions. In April 2001 the committee held a workshop in Washington, D.C., *Academic IP: Effects of University Patenting and Licensing on Commercialization and Research,* examining both the external and the internal effects of the surge in university activity since 1980. In August 2002 the committee conducted a public forum to review the USPTO's 21st Century Strategic Plan, which proposed major changes in patent administration to cope with the lengthening pendency of patent applications and public concerns about examination quality. Speakers at these committee meetings and conferences are listed in Appendix B along with other generous contributors to the committee's deliberations. Audio tracks, slide presentations, and transcripts from the February 2000 and April and October 2001 conferences are available on a CD-ROM, *Patents in the 21st Century,* accompanying this report.

A major challenge in evaluating the patent system is that the effects are specific but the law is general. In some fields, products are protected by a single patent; in others, a number of patents must be acquired or licensed prior to producing and marketing a product to avoid subsequent patent infringement charges that could jeopardize the investment in product development and production facilities. In still other fields, research can be conducted without regard to patents. In some circumstances, brand loyalty, lock-in, and lead time enable producers to recoup costs and make profits; in other cases, these advantages are small, and producers must rely on some other means of protecting investments. There is a tendency to identify problems in the patent system and solutions to them from the perspective of one technology or industry. We try to strike a balance between the specific circumstances and the general patent law. Not everything we say applies to every field, but sensible recommendations depend on taking diverse fields into consideration.

Finally, a word about the title of this report: It is being released soon after the 200th anniversary of the U.S. Patent and Trademark Office and the 50th anniversary of the Patent Act of 1952, but we have no illusion that our recommendations, if adopted, would result in an ideal patent system serving the interests of the American people for 100 years without continuing change. For example, there are important features of the patent system that we did not examine in depth. First, patents exist in most countries, and the degree to which countries at different stages of economic development should adhere to the same standards of patentability, conform to the same rules, and follow the same administrative procedures is an enormously complex although extremely important set of issues. We have confined ourselves to considering the relationships among the U.S., European, and Japanese patent systems not only because they affect the majority of world commerce but also because through diplomacy and by example they influence how other countries' systems are designed. Nevertheless, readers should not infer that what we recommend for the United States we believe less-developed countries should adopt. Second, the training, recruitment, and retention of the examiner corps are obviously relevant to the quality of examination; but the subject exceeded our resources. Third, the fees paid by applicants and patent holders, a subject of intense debate in the context of the proposed Strategic Plan and pending legislation, are a factor in both the transactions costs borne by private parties and the resources available to the government to administer the system. We address the fee structure only in general terms. Fourth, knowing that it was the subject of studies being conducted simultaneously by antitrust experts in the Federal Trade Commission and the Justice Department's Antitrust Division, we decided not to consider in any depth the relationship between competition policy and patent policy. Nevertheless there is a high degree of consistency between the FTC's recommendations and our own. Fifth, although we observe that high damage awards and injunctions in several well-publicized lawsuits since the early 1980s have contributed to the higher importance that firms generally attach to patents, we have not examined the terms of judicial remedies for patent infringement nor the basis on which courts make awards and issue injunctions, nor the role of judges versus juries in patent cases. Although we are aware of much

TABLE 1-1 Principal Recommendations of Panels and Institutions Studying the Patent System

	1919 NRC Patent Committee	1936 NRC Committee on Patents and New Industries	1943 National Patent Planning Committee
USPTO/Examination			
Status	Independent agency		
Fees, resources, and personnel	Increase number of examiners and salaries	Increase number of examiners; annual tax to maintain patents, rising over time	
Evaluation			
Subject matter			
Priority			
Application publication		Publish applications; encourage prior art submissions	
Prior art			
Standards			
Opposition			Considered and rejected
Patent Term			20-year term

1966 Commission on the Patent System	1978 Advisory Committee on Industrial Innovation	1992 Advisory Committee on Patent Law Reform	2003 Federal Trade Commission
Budget adequate for first-class staffing and equipment	Revised fees to support USPTO; maintenance fees	Budget sufficient to achieve 18-month average pendency	Adequate (more) funding
Improved evaluation process and annual quality ratings			
Computer programs not patentable		Computer programs patentable	Consider all costs and benefits in extending to new subject matter
First-to-file with preliminary applications		First-to-file with provisional applications	
Publish applications		Publish applications	Eliminate exception so all applications are published
Recognize foreign art; revise criteria for prior art			Applicant to state relevance of prior art
Applicant to have burden of persuading USPTO			Tighten non-obviousness standard; second review in selected areas
Ex parte pre- and postgrant	Institute a re-examination procedure	Revise re-examination to encourage third party participation	Postgrant opposition
20-year term		20-year term	

continued

TABLE 1-1 Continued

	1919 NRC Patent Committee	1936 NRC Committee on Patents and New Industries	1943 National Patent Planning Committee
Courts/litigation			
Trial		Use technical advisors or juries	
Appellate	Establish court of patent appeals	Establish court of patent appeals	Establish court of patent appeals
Validity			
Infringement/ remedies	Money damages and injunctions		
Licensing		Compulsory licensing rejected	Considered compulsory licensing without recommendation

scholarly discussion of these subjects and some criticism of current practices, they were not raised as problematic in our preliminary conference and roundtable meetings nor in later testimony to the committee. Finally, except as examples of recent legislative changes, we do not consider special purpose statutes such as the Hatch-Waxman Act with patent provisions applying to a single industry. Here there is much controversy about the patenting and patent litigation behavior of both pharma and generic drug companies; but the issues are complex, largely distinguishable from the general working of the patent system, and in any case the statute has recently been modified to address some of the concerns.

1966 Commission on the Patent System	1978 Advisory Committee on Industrial Innovation	1992 Advisory Committee on Patent Law Reform	2003 Federal Trade Commission
Use civil commissioners		Use experts apart from advocates and reconsider jury trials	
	Establish court of patent appeals		
Presume examiner claims rejections are correct		Eliminate best mode; more objective standard of inequitable conduct	Challenge to validity on basis of preponderance of, not clear and convincing, evidence
			Tighten standard of willful infringement

Another reason not to consider our report definitive is that technology and the economy change rapidly, and the patent system needs to adapt, albeit more slowly and gradually. As we assert in Chapter 2, the patent system, along with other policy influences on innovation, should be reviewed periodically to see what adjustments are needed. Our report supports Justice Breyer's belief that this evaluation, although it must rely heavily on the patent bar and other direct stakeholders, should not be confined to them but should include economists, scientists, technologists, and managers making investment decisions. The stakes have grown too high to exclude any relevant expertise.

2

Six Reasons to Pay Attention to the Patent System

INTRODUCTION

For more than half a century the United States has led the world in the development of new technologies and creation of new products. Our international competitive advantage rests in part on the encouragement given to scientific and technological progress by public and private institutions. An open entrepreneurial economy, fueled by effective capital markets and vigorous competition, helps translate these advances into industrial innovation.

This capacity did not appear to be so robust or enduring in the 1970s, when productivity growth rates fell sharply, nor in the 1980s, when Japanese competition fostered the notion that U.S. manufacturing industries were on the decline. But by the mid-1990s the U.S. economy was again exhibiting high productivity growth. A variety of econometric and sectoral studies attributed this robust performance to high rates of innovation, especially in information technologies—semiconductors, computer software, and telecommunications—and their application across the growing service sector of the economy as well as in manufacturing (NRC, 1999a,b; Jorgenson and Stiroh, 2002). In spite of the economic slowdown and the stock market slump in 2001, productivity growth has continued at a rate higher than at any time since 1973. Even through economic cycles, innovation is alive and well in the American economy.

Granting and protecting intellectual property rights are together one of the oldest direct government interventions in the economy and the only policy instrument expressly ordained by the U.S. Constitution to promote innovation. Patents on novel, useful, non-obvious inventions and copyrights on works of literature, art, and other expression are granted on the assumption that although firms and individuals have many incentives to invent and create, some innovations are less likely to be forthcoming in the absence of a grant of exclusive rights providing an opportunity to recoup initial investments while excluding imitators. As a *quid pro quo* for a period of exclusivity, patents, in addition, are assumed to promote innovation by disclosing know-how that might otherwise remain secret.[1]

REASONS THE PATENT SYSTEM MERITS ATTENTION

High levels of innovation in the United States would seem to be evidence that the intellectual property system is working well and does not require fundamental changes. But there are at least six reasons why intellectual property policy has drawn the National Academies' attention and deserves continued scrutiny.

1. The patent system, like other important innovation policy tools, merits periodic examination to help ensure the vitality of the national innovation system.
2. Significant changes in the patent system during the 1980s and 1990s, generally in the direction of extending and strengthening patenting, should be evaluated.
3. The use of the patent system for inventions related to research tools and discoveries has prompted a debate about whether such patents provide incentives to innovate or may in some circumstances impede research progress.
4. Patents are being more actively acquired and vigorously enforced.
5. The roles and benefits of patents vary greatly from one technology or industry to another, but there has been very little systematic investigation of the differences.
6. In the meantime the financial and opportunity costs of acquiring, defending, and challenging patents are increasing.

Preserve America's Capacity to Innovate

The American economy's innovation capacity, although resilient, is not foreordained. To sustain it, all of the chief public policy instruments affecting its vitality deserve periodic examination by analysts as well as stakeholders.

---

[1]Other forms of intellectual property rights—trademarks and trade secrets—do not confer exclusive rights in protected inventions; rather, they are branches of unfair competition law. The federal trademark laws, protecting registered brand names and corporate insignia, operate largely to protect consumers against confusion as to the source of goods and services. Trade secret laws, primarily at the state level, protect against industrial espionage and misuse of confidential business information (Pooley, 1997-1999). Contracts frequently are used by businesses to protect information, and to the extent that they act to define and reinforce the trade secret right, are widely enforced. Other agreements, such as noncompetition covenants and prohibitions against reverse engineering, although often sought in the name of trade secret protection, are more controversial because of their possible effect on fair competition. The Economic Espionage Act of 1996 expanded the effective protection of trade secrets by providing federal criminal penalties for behavior that was traditionally addressed for the most part by state civil law.

In 1999 the National Academies' Board on Science, Technology, and Economic Policy completed an in-depth study of 11 U.S. manufacturing and service industries to determine whether the impression of stronger competitive performance in the 1990s compared to the 1980s was accurate, and if so, what were its sources. The board concluded that the general picture had improved, thanks to a variety of factors, including private sector strategies—firm repositioning, product specialization, consolidation, internationalization of operations, manufacturing process improvements, and cost reduction—driven by vigorous foreign and domestic competition. In addition, the U.S. government followed a supportive mix of macroeconomic and microeconomic policies—deficit reduction, conservative monetary policy, scaling back of economic regulation of transportation, finance, and communications, trade liberalization, relatively permissive antitrust enforcement, and at least until the 1990s, continued support of research across a broad range of scientific and engineering fields. The board observed that none of these favorable conditions was permanent and in some areas—the slowing production of domestic science and engineering talent and the real decline in public support of research in most of the physical science and engineering fields for nearly a decade—the trends were troubling (NRC, 2001).

In one area of public policy—intellectual property rights—the board concluded that evidence of its contribution to the industrial resurgence was lacking. This was not because intellectual property policy was static; on the contrary, it was of one of the most dynamic areas of microeconomic policy in the 1980s and 1990s. Rather, the uncertainty was attributable to the fact that the economic effects of intellectual property policy developments received little study.

This report addresses only the patent system because it affects innovation in more economic sectors than any other form of intellectual property protection and because copyright policy, at least in the context of digital media, has been the subject of recent National Academies study (NRC, 2000). By contrast, the National Academies have not examined the patent system broadly since 1936, and until the Federal Trade Commission issued a report of the inquiry that was conducted in parallel with our investigation it had been more than a decade since the last government review of patent policy.

Since the Patent Act of 1952, the last comprehensive restatement of patent law, three government-appointed panels have deliberated and made legislative and administrative recommendations. The most recent, the Advisory Commission on Patent Law Reform, was appointed by Commerce Secretary Robert Mosbacher in the first Bush administration and reported to his successor, Barbara Franklin, in 1992. Earlier panels were the presidentially appointed Commission on the Patent System, reporting to President Lyndon Johnson in 1966, and the patent policy subcommittee of the Advisory Committee on Industrial Innovation, a multi-agency policy review at the end of the Carter administration. Each of these committees was composed almost entirely of senior managers of Fortune 100 companies, in-house patent counsels of such firms, and members of law firms with large corporate clients. They included no economists or other social scientists, legal scholars, active scientists or engineers, independent inventors, investors in technology-based firms, or people with recent experience in the judicial branch of government.

Despite its utilitarian economic rationale and bearing on the progress of science and technology, patent policy has never been an integral element of either economic or science and technology policy making. Much attention has been focused on other countries' conformity with contemporary U.S. standards of intellectual property protection as an aspect of trade policy, on the allocation of intellectual property rights to the results of publicly funded research as an aspect of research and development (R&D) policy, and on the exercise of intellectual property rights as an aspect of antitrust enforcement. Patent policy *per se,* nevertheless, has not been on the agendas of the Council of Economic Advisers, National Economic Council, or commerce and science committees and subcommittees of Congress. Rather, it has been the preserve of practitioners, corporate stakeholders, the U.S. Patent and Trademark Office (USPTO), Senate and House Judiciary Committees, and the federal appellate courts.

As the introduction to the collected papers commissioned for this project suggests (Cohen and Merrill, 2003), there has been a blossoming of empirical research on and theoretical analysis of the functioning of the patent system during the past 15 years. Although this literature falls far short of providing a definitive answer to the general question, "Are patents doing their job in the information economy?" it is beginning to describe the role patents play in important industrial sectors and to assess the effects of policy changes implemented during this period.

In short, a study drawing upon a wider range of expertise and experience is timely. Domestically, the fact that the innovation system of which intellectual property policies are a part is working well by historical and international comparative standards suggests that the patent system is not broken. But there may be instances in which the use of patents is not working as well to promote innovation as we might hope and well-thought-out changes could improve the system's functioning. Moreover, what the United States does with respect to intellectual property policy influences what other countries do, both through negotiation and by example.

Substantial Changes in Patent Policy

The patent policy landscape has changed significantly in the last 20 years, and the consequences have not been examined systematically.

At the end of the 1970s the patent system was widely perceived to be weak and ineffective, unable to keep up with fast-moving technological changes, under attack by antitrust authorities, and of only limited value to patent holders. Patenting by U.S. inventors residing in the United States was constant or declining through the 1970s (Jaffe, 2000). Beginning in 1980 a series of legislative actions, judicial decisions, executive branch initiatives, and international agreements largely spearheaded by the United States[2] ostensibly strengthened the rights of intellectual property owners and extended intellectual property rights (IPRs) into new areas of technology.[3] This policy thrust at the national level has extended to other forms of intellectual property protection. For example, the Digital Millennium Copyright Act (DMCA) of 1998 strengthened the protection of material in digital form, while the Sonny Bono Copyright Extension Act of 1998 lengthened copyright terms from 50 years to 75 years beyond the lifetime of the creator. The 1996 Economic Espionage Act subjected some trade secret misappropriation to federal criminal penalties, whereas previously it had been a matter of state civil law. And the Trademark Dilution Act of 1995 extended the rights of mark holders beyond the avoidance of consumer confusion. Most important from our perspective, many of the IP policy changes involved the patent system. These changes can be classified as steps to (1) extend patenting to new subject matter; (2) strengthen the position of patent holders vis-à-vis infringers; (3) encourage new classes of patentees; (4) extend the duration of some patents; and (5) relax antitrust limitations on the use of patents.

New Technologies

- In *Diamond v. Chakrabarty* (1980)[4] the Supreme Court confirmed the eligibility for patenting of organisms with artificially engineered genetic characteristics. Thereafter the USPTO granted innumerable biological material as well as biotechnology final product patents.
- The Supreme Court in a 1981 decision, *Diamond v. Diehr,*[5] upheld the patentability of inventions incorporating a computer software program as an adjunct to a physical process, ushering in an era in which software is commonly protected under both copyright and patent law.
- The Court of Appeals for the Federal Circuit ("Federal Circuit") in a 1998 case, *State Street Bank & Trust Co. v. Signature Financial Group,*[6] upheld the

---

[2]An exception is the *sui generis* protection of databases adopted by the European Union in 1996. So far, Congress has not adopted the European system (http://www.arl.org/info/frn/copy/iff1.html).

[3]From a legal perspective it may be more accurate to characterize the court decisions addressing the patentability of genetically modified organisms, software, and business methods as confirming the patentability of all technologies rather than as extensions of patenting. The latter term reflects common understanding, however.

[4]*Diamond v. Chakrabarty*, 447 U.S. 303, 100 S. Ct. 2204, 65 L. Ed. 3d 144, *available at* 1980 U.S. LEXIS 112, 206 U.S.P.Q. (BNA) 193 (1980). Although now taken for granted, the case was decided on a 5-4 vote.

[5]*Diamond v. Diehr*, 450 U.S. 175, 101 S. Ct. 1048, 67 L. Ed. 2d 155, *available at* 1981 U.S. LEXIS 73, 209 U.S.P.Q. (BNA) 1 (1981).

[6]*State St. Bank & Trust Co. v. Signature Fin. Group*, 149 F.3d 1368, *available at* 1998 U.S. App. LEXIS 16869, 47 U.S.P.Q.2d (BNA) 1596 (Fed. Cir. 1998).

patentability of methods of doing business as well as that of software, so long as in either case the invention is expressed as a method that accomplishes useful, concrete, tangible results.

Strengthening Patent Holders Vis-à-Vis Alleged Infringers

- In 1982 Congress established the Federal Circuit to handle, among other matters, patent litigation appeals from the federal district courts and appeals from decisions of the Board of Patent Appeals and Interferences (BPAI), the administrative law body of the USPTO. As a result, the wide variation in circuit appeals courts' treatment of patent infringement cases was sharply curtailed, generally to the benefit of patent holders. The success rate of plaintiffs (that is, findings that a patent was valid and infringed) in appeals increased significantly as a result of court reform.[7]
- In the same period, plaintiffs' damage recoveries in a handful of highly visible patent suits had a significant demonstration effect. For example, in its suit against Kodak for infringement of instant camera patents, Polaroid was awarded nearly $900 million and Kodak was ordered to cease production.
- The 1988 Process Patent Amendments Act[8] enabled U.S. process patent holders to block the import of foreign products produced by methods infringing their patents as well as to hold domestic sellers or users of a product made by a patented process liable for infringement.
- As part of the Uruguay Round trade negotiations under the General Agreement on Tariffs and Trade (GATT), the Trade-Related Aspects of Intellectual Property Rights (TRIPS) Agreement was concluded in 1994. It requires World Trade Organization (WTO) members to protect most commercially important technologies and limits their ability to compel the licensing of patents. In addition to multilateral negotiations, the United States pursued strong IPR protection in a series of bilateral and regional venues in the 1980s and 1990s and continues to do so.
- Until very recently it was widely believed that purely research uses of patent inventions were shielded from infringement liability by an experimental use exception first articulated in 19th century case law. But in *Madey v. Duke University,*[9] a suit brought against the university by a former professor and laboratory director, the Federal Circuit dispelled that notion by holding that there is

---

[7]In a comparison of appeals cases from 1953 to 1978 and from 1982 to 1990, the share of District Court decisions finding validity and infringement that were upheld increased from 62 percent to 90 percent. Decisions of invalidity and no infringement were reversed 12 percent of the time before the Federal Circuit's creation and 18 percent afterward. Moreover, the rate of preliminary injunctions increased dramatically. See Lerner (1995); Lanjouw and Lerner (1997); Allison and Lemley (1998); and Jaffe (2000).

[8]35 U.S.C. § 271(g).

[9]*Madey v. Duke Univ.* 307 F.3d 1351, *available at* 2002 U.S. App. LEXIS 20823, 64 U.S.P.Q.2d (BNA) 1737 (Fed. Cir. 2002).

no protection for research conducted as part of the university's normal "business" of investigation and education, regardless of its commercial or noncommercial character.

New Patent Holders

- The Bayh-Dole Patent and Trademark Amendments Act of 1980[10] made it the general rule that universities, other nonprofit institutions, and small businesses could acquire exclusive rights to inventions developed with federal support. Partly as a result, patenting by universities soared although their share of the total remains very small. Gradually, this policy was extended to all federal contractors and research grantees with narrow exceptions. The Stevenson-Wydler Act of the same year gave federal research agencies and their investigators additional encouragement to patent and license the results of in-house research.

Extended Patent Terms[11]

- The 1984 Drug Price Competition and Patent Restoration (Hatch-Waxman) Act,[12] while exempting from infringement regulatory testing of generic pharmaceuticals, allowed patent term extensions on new drugs of up to five years if the drug's approval is subject to regulatory delay.

Relaxed Antitrust Limitations on the Use of Patents

- From the 1980s onward there was a marked evolution in the attitude of the Justice Department's Antitrust Division and the Federal Trade Commission toward business conduct involving patents, resulting in a much more nuanced and pro-patent position (FTC, 2003). In 1981 the division's deputy assistant attorney general abandoned a list of nine licensing practices that the department a decade earlier had characterized as automatically illegal.
- The 1988 Justice Department *Antitrust Enforcement Guidelines for International Operations* outlined the consumer benefits from intellectual property licensing and adopted a rule-of-reason approach to such issues.

---

[10]35 U.S.C. § 200 et. Seq.

[11]To comply with TRIPS, U.S. legislation changed the life of U.S. patents from 17 years from date of issue to 20 years from date of application. In practice this is an extension of patent life only in cases where application pendency does not exceed 3 years. On the other hand, the length of the period for which a patent holder may collect damages from an infringer *has* been extended from 17 years to 18 years, the period between publication of the patent's application and its expiration.

[12]Drug Price Competition and Patent Term Restoration Act of 1984, Pub. L. No. 98-417, 98 Stat. 1585 (codified at 15 U.S.C. §§ 68b-68c, 70b (1994); 21 U.S.C. §§ 301 note, 355, 360cc (1994); 28 U.S.C. § 2201 (1994); 35 U.S.C. §§ 156,271,282 (1994)).

- In 1995 the Justice Department and the Federal Trade Commission jointly issued *Antitrust Guidelines for the Licensing of Intellectual Property,* reiterating the 1988 principles and declaring that "the Agencies do not presume that intellectual property creates market power in the antitrust context" and intellectual property licensing is "generally procompetitive."

Some of the reasons for this unidirectional "ratcheting up" of patent rights are apparent—a general belief in the efficacy of the intellectual property system and a reluctance to disrupt reasonable investment-backed expectations once created by law or regulation. Strictly speaking, whether the changes contributed to a "strengthening" of patent rights is debatable. Some argue that a lowering of the threshold conditions of patenting, especially the standards of utility and nonobviousness, has led to the issuance of large numbers of "weak" patents unlikely to stand up in litigation. Others have defined "strength" as a function of the breadth of individual claims in issued patents (as well as the range of patentable subject matter, the duration of patents, and the likelihood that granted claims will be enforced in court against infringement or invalidity challenges) and point out that recent decisions of the Federal Circuit have forced applicants and examiners to narrow and possibly proliferate patent claims (Gallini, 2001). What is not debatable is the marked turnaround in public policy that has led to the apt characterization of the last 20 years as a "pro-patent era" (Cohen, 2002).

The effects of some of these actions were only beginning to play out when the Mosbacher Commission reported 10 years ago, and other significant changes lay ahead. The patent system is always evolving, and the effects of these changes take a considerable period of time to be felt. In the meantime it is important to ask several questions: What, so far as we can tell, have been the costs and benefits of the actions taken in the last several years and the consequences intended or not? What should be the direction of patent policy in the next decade and beyond? Should we continue to extend patenting and patent rights or modify that course?

Expanded Patenting of Research Tools and Discoveries

There is disagreement whether patents on discoveries and tools of research, an expanded domain of patenting,[13] *provide needed incentives to innovate or, because of difficulties and costs entailed in accessing the subjects of these patents, may impede the progress of scientific investigation.*

Advances in most technologies are cumulative, that is, they build upon one another. As a result, how exclusive rights to a pioneering invention affect follow-on innovation has always been an issue for theorists and occasionally historians

[13]The reference is primarily to biological material, which is difficult to invent around, not to laboratory equipment, which has been patented for some time.

and policy makers (Merges and Nelson, 1990; Scotchmer, 1991). This influence is a function of both the scope of patent claims allowed and the behavior of patent owners. For a few notable commercial product inventions—Edison's incandescent lamp[14] and the Wright brothers' airplane stabilization and steering system[15]—broad pioneering patents were exercised in a manner that at least temporarily deterred competitors from making further improvements. The patent holders either aggressively enforced their rights or refused to enter into licensing agreements. Radio illustrates the possibility that when separate patent holders with broad enabling patents (in this case, Marconi Company, De Forest, and De Forest's main licensee, AT&T) cannot agree on licensing terms, technological progress may be impeded for a time. Eventually, in all of these cases the obstacles were overcome by industry consolidation or government intervention in or near wartime to compel licensing or patent pooling (Merges and Nelson, 1990; Merges, 1994).

The issue has recently reemerged in a new context—not whether failure to license or cross-license product patents is impeding further innovation but whether patents on some research tools and foundational discoveries have the potential to stymie further scientific research well upstream of commercial products (Nelson, 2003). The concern involves a rapidly expanded domain of patenting—inventions that are useful solely or primarily for further research. Previously, in most cases these techniques and discoveries became part of the public domain of scientific knowledge available without restriction for use by all investigators, especially where they were the products of publicly funded research at institutions of higher education. Open academic science thrived not on the basis of altruism but because the rewards for successful work are reputational and the benefits that go with prestige. That they are now being patented may be as much a function of changes in the innovation system as of the utilization of patents in new fields of technology.

Underlying the concern is the presumption that the payoffs not only of the most fundamental scientific research but also of research directed at solving practical problems are frequently serendipitous, and the chances of progress are greatest not only when scientists are free to attack what they see as the most challenging scientific problems in the ways they think most promising but also when competing approaches are in fact pursued (Bush, 1945). A closely associated belief is that scientific progress requires that research results be open for all to use, attempt to replicate, and evaluate (Merton, 1973).

Three different problematic circumstances have been hypothesized:

1. Access to patented foundational discoveries is denied, foreclosing research avenues to other investigators (Merges and Nelson, 1990).

---

[14]U.S. Patent No. 223,898.
[15]U.S. Patent No. 821,393.

2. Access to patented discoveries or research tools is possible but on terms that make their use too costly, at least for nonprofit research performers.

3. Pursuit of research is effectively blocked because of the practical difficulty of acquiring rights to use all of the needed patented elements of research held by diverse parties (Heller and Eisenberg, 1998).

The concern has focused primarily on the field of biotechnology, where there has been an increase in patents on a variety of inputs into the process of discovering a drug or other medical therapy or method of diagnosing disease as well as the tools of plant modification—genes and genetic sequences, drug targets and pathways, antibodies, and so forth. There is no ostensible reason why concern about the impact of patents on science will be confined to biomedical research, but it is easy to understand why the foreclosing or restricting of opportunities to develop better medical therapies and diagnostics is alarming to some. Moreover, with respect to biotechnology, where many of the patents are on naturally occurring substances, albeit ones that are isolated and purified, there may be fewer opportunities to avoid patent infringement by "inventing around" existing claims than there are in other fields.

This set of concerns is by no means universal, and not all members of this committee share it. Historically, the existence of blocking patents is exceptional, and although breakdowns in negotiations occur, rights over essential inputs to innovation are routinely transferred and cross-licensed in industries, such as semiconductors and communications, where there are numerous patents associated with a product and multiple claimants (Levin, 1982; Hall and Ziedonis, 2001; Cohen et al., 2000). The Moving Picture Experts Group (MPEG) Consortium is a recent case in point.[16] In Japan, where across the manufacturing sector there are many more patents per product than in the United States, licensing and cross-licensing are commonplace (Cohen et al., 2002). This is likely to be the case with most research tool patents, which are of little or no value unless the tools are used widely.

Nor can it be assumed that patents on genes, genetic sequences, proteins, and other natural substances effectively preclude circumvention. Generally speaking, diseases result from a variety of mechanisms rather than a single mechanism and can be treated using different pathways. Competing patented pharmaceutical inventions—for example, Viagra and Levitra, Previcid and Nexium—can have similar biological effects. Thus, at least in some cases, the established method of circumventing a monopoly patent position applies in biomedicine and in agricultural biotechnology as it does in other fields. The incentive to make the effort depends on the market prospects.

---

[16]See http://www.mpegla.com.

Finally, it is argued that if many "upstream" innovations have become sufficiently valuable to patent, their development in some cases may depend upon the patent system's incentives. Although it is likely that most research tools are created simply to facilitate a research objective or to overcome obstacles, it may be that some valuable tools would not be invented without the incentive of exclusivity.

Even if there were problems of impediments to research, there would not be agreement on its sources or remedies. Some observers believe that some research tool patents have crossed over into traditionally unpatentable subject matter—scientific facts or principles or natural phenomena with negligible human intervention. Others believe the issue is one of unreasonably low standards of utility or non-obviousness, or excessive patent scope that allowed claims on some research tool patents covering more than the described invention and its application. Still others are of the view that the problem would not exist or would be manageable if noncommercial research activities were shielded from patent infringement liability. Nonetheless, it is the conflicting factual claims that merit first attention, and we address them in the next chapter.

Surge in Patent-Related Activity

Patents are being more frequently acquired and vigorously asserted and enforced. The surge in patent-related activity is indicative that firms in a variety of businesses as well as universities and public entities attach greater importance to patents and are willing to incur higher costs to acquire, exercise, and defend them.

The number of U.S. patents issued to both U.S. and foreign entities nearly tripled from 66,290 in 1980 to 184,172 in 2001.[17] If patenting by U.S. entities is calibrated by domestic population growth or R&D spending, the increase is less but still significant, especially in the 1990s (see Figures 2-1 and 2-2). Patents per million dollars of R&D rose about 47 percent from 1985 to 1997, increasing from 0.18 patents per million dollars to 0.34 patents per million dollars. An exception is pharmaceuticals whose R&D investment growth has exceeded its patenting rate.

Economists who have studied the phenomenon are not in complete agreement about the causes of the patenting surge, but most give a good deal of credit to the policy changes in the 1980s and 1990s, especially the creation of the Federal Circuit and the resulting higher rates at which patent validity and patent holders prevailed in litigation (Kortum and Lerner, 1998; Hall and Ziedonis, 2001). In addition, the increasing competitiveness of national and global markets has no

[17]The increase was 200 percent for foreign entities versus 150 percent for U.S. inventors, but economic growth and R&D expenditure increases were greater abroad than in the United States during much of this period.

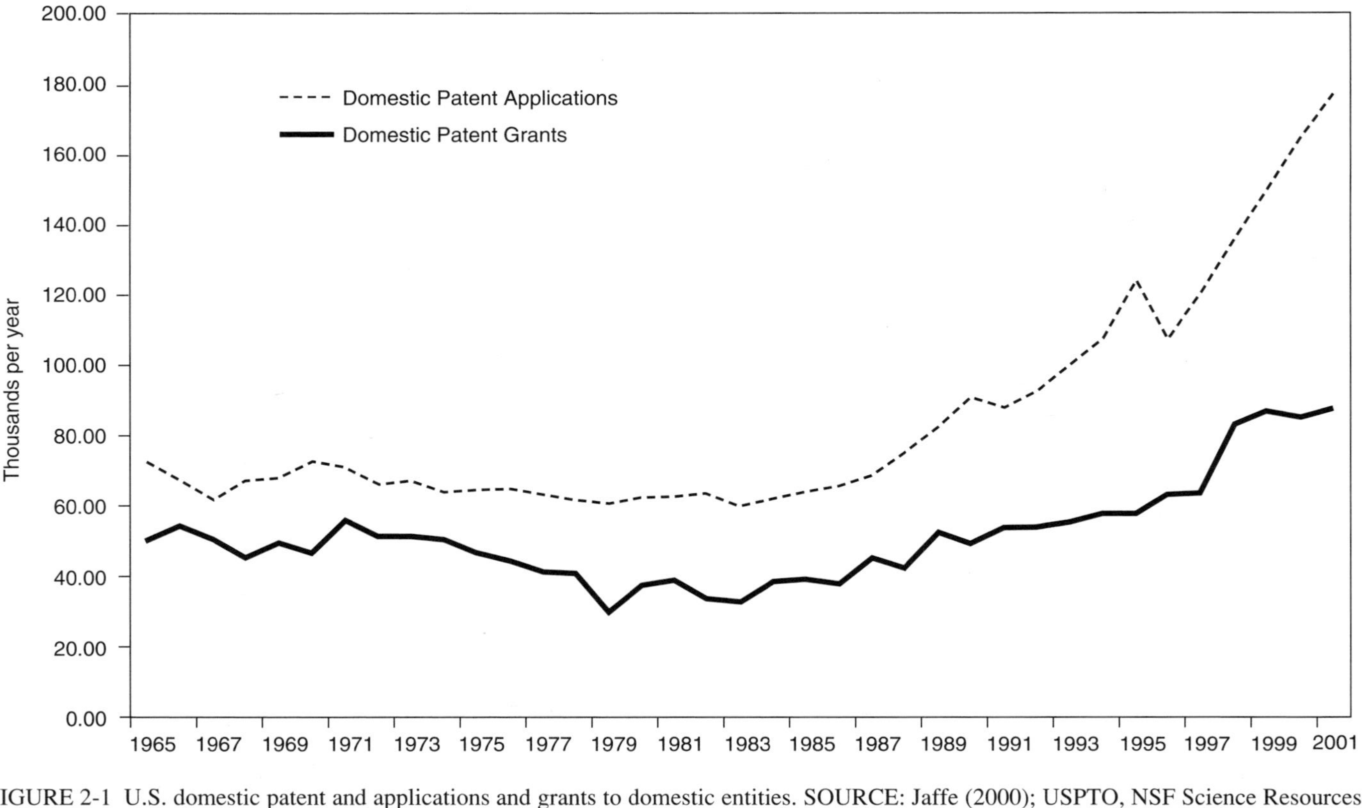

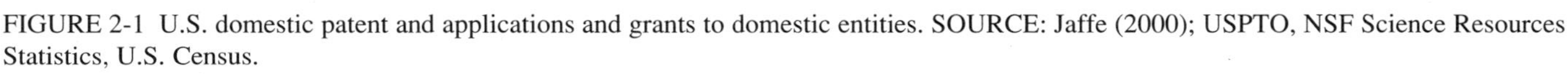
FIGURE 2-1 U.S. domestic patent and applications and grants to domestic entities. SOURCE: Jaffe (2000); USPTO, NSF Science Resources Statistics, U.S. Census.

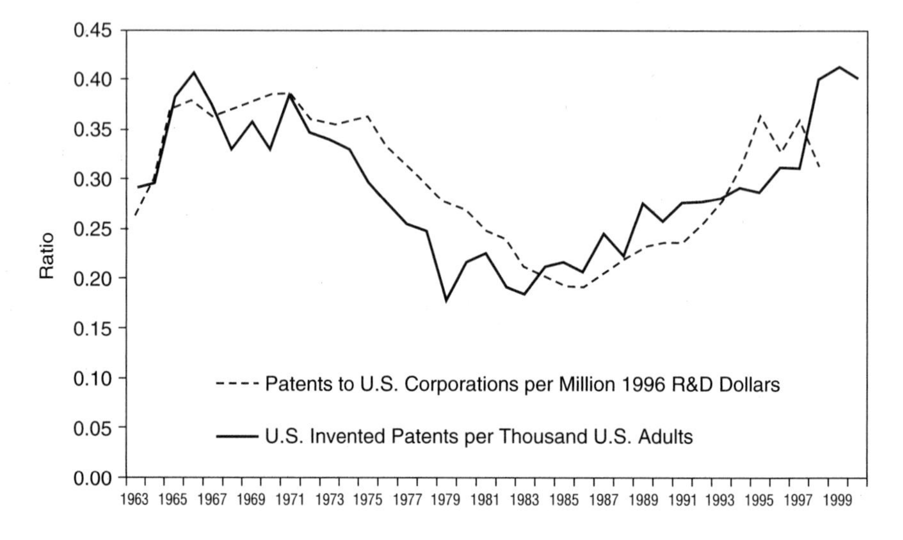

FIGURE 2-2 Patent and corporate R&D and population ratios. SOURCE: Jaffe (2000), USPTO, NSF Science Resources Statistics, U.S. Census.

doubt encouraged firms to exploit new ways of protecting market positions, especially since economic regulation, trade barriers, and artificial monopolies have been reduced. A case in point is the telecommunications industry (Bekkers et al., 2002).

The growth is not distributed evenly across technology areas or industries, however. The number of patents per R&D dollar, used by many as a measure of the "patent propensity" of firms, increased by about 50 percent for U.S. corporations during the 1985-1998 period. But Hicks and colleagues' findings (2001), although not spanning precisely the same period, suggest that information technology (IT) may account for much of this increase. IT patents per R&D dollar increased from an average of 0.28 patents per million dollars to 0.48 patents per million dollars between the periods 1989-1992 and 1993-1996.[18] In contrast, in health, chemical and polymer, and all other technologies, the patent propensity over the same period changed, respectively, from 0.23 to 0.24, 0.38 to 0.38 (no change), and 0.37 to 0.35 (a decline). If patents are classified by industry rather than technology, the IT sector also accounts for most of the growth in patenting (Hall, 2003a). This is reflected in any recent list of companies receiving the most U.S. patents. In 2002, eight of the top ten companies receiving U.S. patents were predominantly in IT; five of those were Japanese-headquartered.

[18]The associated patent counts here drawn from the periods 1991-1994 and 1995-1998, respectively.

TABLE 2-1 Higher Propensity to Keep Patents Valid

Patent Renewal Rates (%)	FY1997	FY1998	FY1999	FY2000	FY2001	FY2002
First stage (end of 3rd yr)	80.3	81.8	83.1	84.3	84.5	85.1
Second stage (end of 7th yr)	55.8	56.6	57.9	59.4	59.9	59.5
Third stage (end of 11th yr)	35.4	36.1	37.7	38.8	39.1	38.4

SOURCE: USPTO, FY 2000 and FY 2002 USPTO annual reports.

On the other hand, all types of firms, not just existing players in the patenting arena, contributed to the increase in patenting. In fact, the share of patents going to small firms and the share going to firms with few previous patents have both increased in recent years (Jaffe, 2000). Likewise, the share of patents issued to universities and government laboratories increased in the 1980s and 1990s. University patents per dollar of research spending more than tripled from 1980 to 1997; the patent propensity of federal laboratories was on a similar course until 1993, when R&D spending in areas other than health started to decline.

Since the 1980s patent holders have been required to pay maintenance fees at the end of the third year, seventh year, and eleventh year to continue to be able to enforce their patents. A large majority of patents are renewed at the first stage, but nearly one-half are allowed to expire at the second stage, and up to two-thirds lapse at the end of the third stage. Nevertheless, the proportion of patents that are renewed has been increasing at all stages in recent years (see Table 2-1).

Unfortunately, there are no aggregate data on patent-related licensing transactions although a few firms have reported rapid growth in licensing revenue, depending on business cycle conditions. IBM's licensing revenue peaked at more than $1.6 billion in 2000 (Berman, 2002). Lucent Technologies' patent portfolio yielded $500 million in 2000.[19] Texas Instruments has pursued a litigation-based strategy. Patented technology is increasingly perceived as having more strategic importance than previously as reflected in the creation of intellectual property practices by nearly all large consulting firms, the emergence of specialized firms that analyze clients' patent holdings and counsel them on using patent portfolios to obtain licensing revenue, the advent of venture-backed firms that purchase unexploited patents and assert them, the use of patent information to pinpoint strategic trends and stock investment opportunities, and the appearance of business management commentary on the importance of a firm's identifying lucrative licensing prospects among its latent patents (Rivette and Kline, 2000). This is, of course, consistent with the frequent observation that many forms of intangible assets—workforce caliber, R&D, brands, and distinctive competences as well as

[19]Source: Daniel McCurdy, former president, Intellectual Property Business, Lucent Technologies.

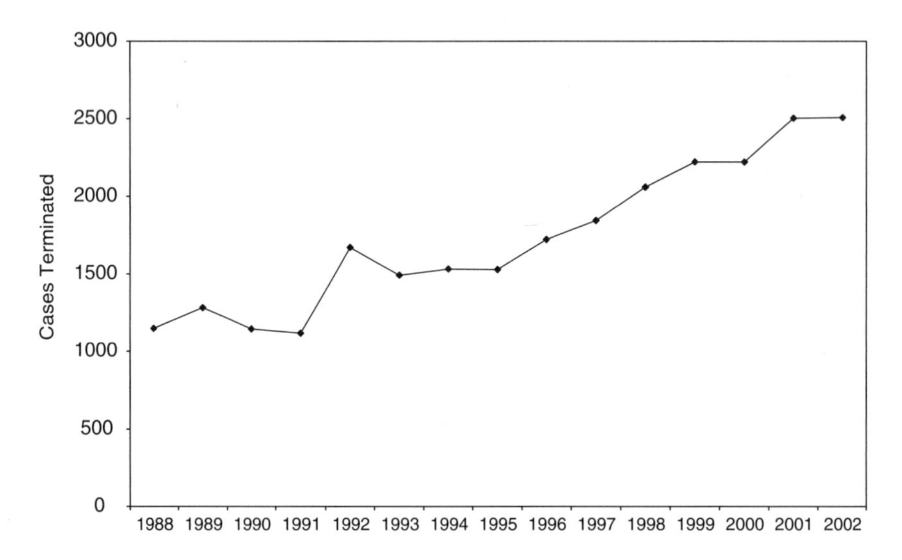

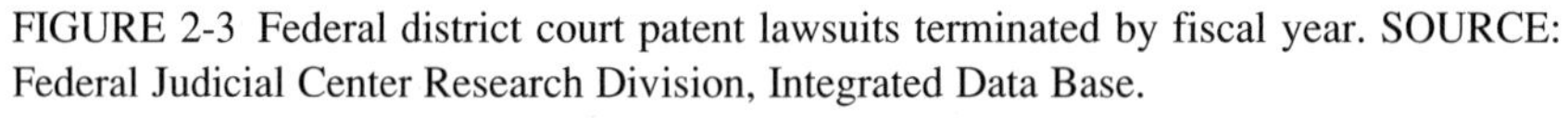

FIGURE 2-3 Federal district court patent lawsuits terminated by fiscal year. SOURCE: Federal Judicial Center Research Division, Integrated Data Base.
NOTE: "Terminated" includes judgments, dismissals, settlements, transfers, and remands.

intellectual property—have increased in value relative to plant and equipment assets.

Another area of rapid growth in patent activity is litigation and legal representation, the latter in all formal processes involving patents (i.e., patent prosecutions, licensing, and litigation). The number of patent lawsuits settled in or disposed by federal district courts doubled between 1988 and 2001, from 1,200 to nearly 2,400 (see Figure 2-3).[20]

The number of practitioners specializing in intellectual property law and affiliating with the American Bar Association (ABA) Intellectual Property Section increased 39 percent between 1996 and 2002 while the ABA membership overall grew 6 percent over the same period (see Figure 2-4).

Many companies rely less on patents than on other means such as marketing, lead time, production and distribution efficiencies, secrecy, and complementary services to achieve market advantages. That is particularly true from the perspective of R&D managers, who, with the notable exceptions of those in pharmaceuticals, chemicals, and medical equipment, have in a series of surveys ranked patents fairly low as a means of protecting inventions and exploiting inventions

[20]The patent litigation rate has not changed; the doubling is due to the increase in patents (Lanjouw and Schankerman, 2003).

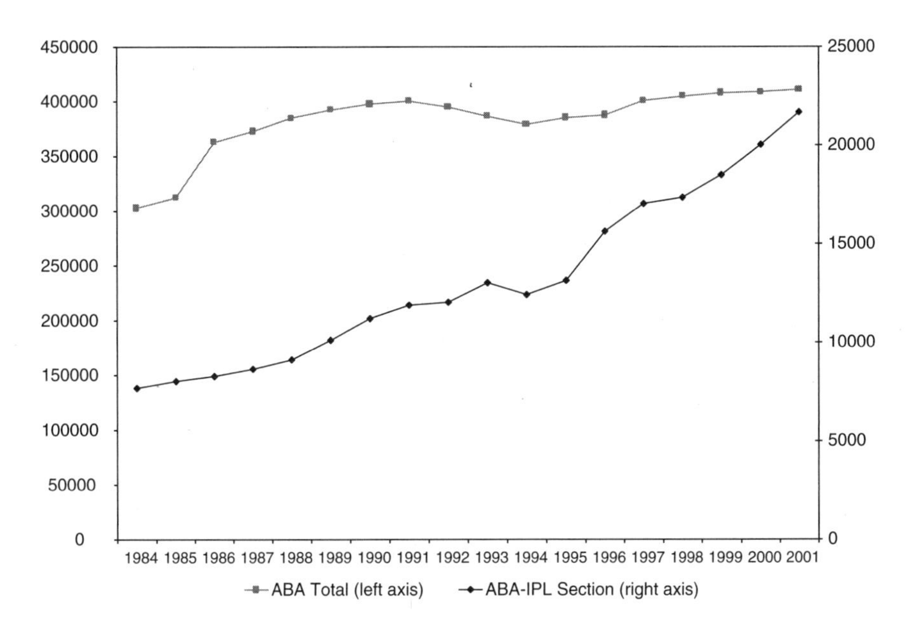

FIGURE 2-4 American Bar Association membership: Intellectual Property Law Section and total. SOURCE: American Bar Association.

(Scherer et al., 1959; Taylor and Silbertson, 1973; Mansfield, 1986; Levin et al., 1987; and Cohen et al., 2000). But among R&D executives of large firms there has been a modest increase in the importance attached to patents between the so-called Yale survey conducted in 1983 by Levin and colleagues and the 1994 Carnegie-Mellon survey (CMS) conducted by Cohen and colleagues. For the protection of product innovation, patents were ranked first or second in 7 of the 33 industries in the Yale survey and in 12 industries in the CMS survey (Cohen, 2000).

This is consistent with other indirect evidence that patents have come to occupy a more central role in corporate decision making. Allison and Lemley (2002) compared a random sample of 1,000 patents issued between 1996 and 1998 with a similar random sample issued 20 years earlier (1976-1978) to determine how the patent system changed over time. Two dramatic changes emerged from the data. First, obtaining a patent has become a more complex process, involving more claims, citing more prior art, taking longer, and involving more refilings. Second, patents today are much more heterogeneous than their counterparts two decades ago. Allison and Lemley suspect that changes in technology and prior art search methods (for example, automated searches of scientific and technical literature) account in part for the changes, but the increased salience of

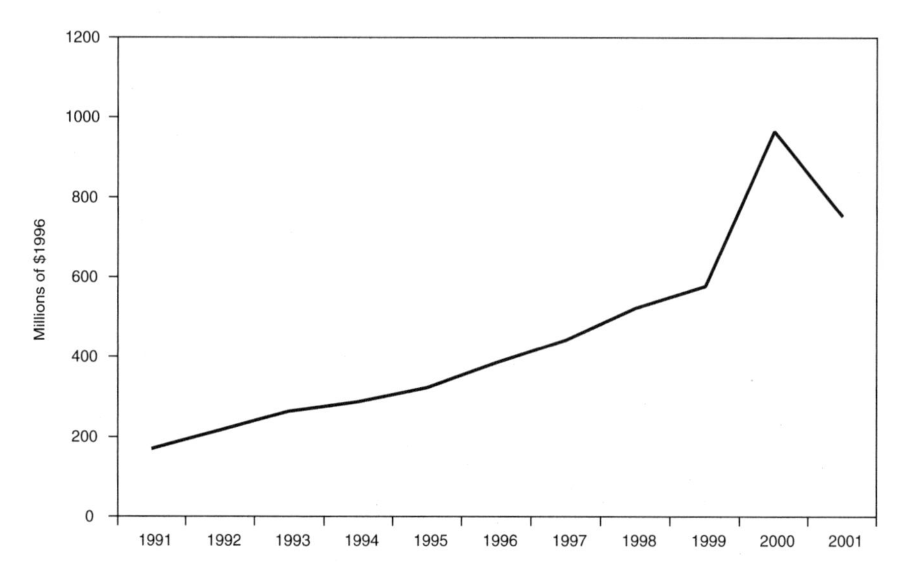

FIGURE 2-5 Adjusted gross licensing income of U.S. research universities. SOURCE: AUTM (2003).

patents to U.S. business may offer a broader explanation of the findings. An increase in the perceived importance of patents has led patentees to invest more in the process of application and examination—asserting more claims, citing more prior art, more frequently amending and refiling applications, tolerating the longer time the examination takes, and even seeking to have their issued patents re-examined when previously unknown prior art comes to light,[21] presumably in order to enhance the eventual patent's value in licensing and litigation.

Perhaps the clearest instance in which the increase in patent-related activity is associated with perceptions of the increasing value of patents is in higher education. Beginning in 1991, university licensing revenue, chiefly from patents, increased nearly three times, from $200 million to $550 million in less than a decade (see Figure 2-5).

Closer examination reveals that a large majority of this revenue derives from a relatively few biomedical inventions and flows to a handful of institutions whose receipts significantly outpace the expenses of patenting, technology transfer, and litigation. The top 10 university patent holders accounted for 66 percent of licensing revenue in 2000 (AUTM, 2003). Nevertheless, the uncertain odds of pay-off have not deterred research institutions from investing heavily in such

---

[21]See Appendix A, *A Patent Primer.*

operations. In 1980, 24 universities reported having technology transfer offices. By 2000 nearly all research institutions had them.

Varied Roles and Uncertain Benefits of Patents

The benefits of more patents in encouraging research and development and simulating innovation appear to be highly variable across technologies and industries and, conceivably, over time; but the industry-specific and comparative research is inadequate to determine the extent of the benefits and the circumstances in which they apply. In many cases patenting activity has departed from its traditional role and has become strategic. Some firms are building large patent portfolios to gain access to others' technologies and reduce their vulnerability to infringement litigation. This may not be a new phenomenon, but the number of players and the number of patents needed to pursue a defensive strategy have almost certainly increased

The traditional rationale for patent protection is to increase the incentive to invent by conferring the right to exclude others from making, using, or selling the invention in exchange for foregoing secrecy by publishing the invention, making the information available for others to build upon. It is often assumed that in a highly competitive environment firms will not invest as substantially in the development of new products and processes without the ability to protect their advances from imitation. But there are theoretical reasons to question how substantial the incentive of patenting is and how broadly the incentive operates across industries. The cost of disclosing the details of one's innovation to competitors through patent publication may be greater than the gain from patenting (Horstmann et al., 1985).[22] The competitive position of rival firms patenting in the same technological domain may be even more enhanced by extensive patenting (Gallini, 2002). And, as we have described, where innovation is cumulative, it matters how and to whom intellectual property rights are first allocated. Subsequent inventors and their incentives and disincentives for research and innovation are affected by the willingness of early patent holders to license each other in instances where inventing around the patents would be difficult. Thus, where innovators are followers, increasing patent strength could increase or it could reduce their incentives to innovate (Merges and Nelson, 1990; Scotchmer, 1991, 1996; Green and Scotchmer, 1995; O'Donoghue et al., 1998; and Gallini, 1992).

In the only empirical study to date attempting to determine a causal link between patenting and R&D, Arora and colleagues (2002) attempt to estimate the additional payoff attributable to patenting an invention relative to the payoff of

---

[22]The cost of disclosure, like the benefits of patenting, probably varies among technologies. Notwithstanding enablement and written description requirements (see Appendix A), some argue that software innovators' disclosures mean little without a requirement to reveal source code in a patent application.

not patenting it and to link that difference to R&D investment in a number of manufacturing industries. Although as with all models the authors use simplifying assumptions that may be questioned, they nonetheless take into account that the appropriability incentive of patenting and R&D decisions are both driven by many of the same factors. The model also considers the role of patents in promoting R&D spillovers, the R&D efficiency gain from the information disclosed in patents, and that if one firm benefits from stronger patents in its line of business so in all likelihood will its competitors. They find that patents have the greatest positive effect on R&D spending in pharmaceuticals, biotechnology, medical instruments, and computers. In semiconductors and communications equipment the incentive effect is much lower, although still positive and not negligible. Although representing an important advance on previous research, the study is not a comprehensive analysis of the social welfare effects of patents. For example, it does not consider the positive or negative impact of patent use on industry entry, which could have an important bearing on innovation. Finally, although the analysis is sensitive to the possible substitution of other appropriability mechanisms (such as secrecy or lead-time advantages) for patents at the margin, it cannot project the possible impact on innovation of eliminating patents altogether.

In the nonmanufacturing part of the economy, it is less clear that patents induce additional investment, for example, in software advances and business method improvements. Possibly as a result, in part, of trade secrecy and copyright protection, invention flourished in both fields well before the advent of patent protection, and open source software development continues under a different incentive system (von Hippel, 2001). Bessen and Maskin (2000) argue that the advent of software patents ushered in a period of stagnant, if not declining, research and development, but they produce no evidence of a direct link between the two phenomena. The fact is that the role and impact of patents in the service industries and service functions of the manufacturing economy have not been studied systematically.

The quid pro quo for giving the patent holder the right to exclude others is to compel disclosure of the invention in terms that enable others to replicate, modify, and circumvent it. Conceivably, the surge in patenting over the past 20 years has resulted in the publication of a great deal of technology that otherwise might have remained secret, and its disclosure might have enhanced the productivity and efficiency of the research and development process. Cohen and colleagues (2002) cast some doubt on this hypothesis, reporting that U.S. corporate R&D managers, relative to their Japanese counterparts, consider patents a much less important source of intelligence on the R&D activities of their rivals than other sources of information, such as publication or technical meetings. The U.S. survey findings are similar to the results of the European Community Innovation Surveys in which firms rank customers, exhibits, conferences, journals, suppliers, competitors, and nonprofit institutions ahead of patent disclosures as technical information sources (Arundel et al., 2002). On the other hand, Cohen and colleagues acknowledge the possibility that in industries such as pharmaceuticals and biotechnology firms allow or encourage their R&D personnel to publish or present results to scientific meetings once patents have been applied for. In those cases the patent system plays a positive indirect role in information diffusion.

The "pro-patent" era is unquestionably associated with a rapid growth in the markets for new knowledge. On the basis of secondary data, Arora and colleagues (2001) roughly estimate that the value of technology licensing in the United States increased from $24 billion in 1990 to $44 billion in 1995 in constant dollars, and the number of deals increased from more than 200 to well over 2,100. Anand and Khanna (2000) support the hypothesis of a direct relationship by showing that licensing is more frequent in industries where patents are also prevalent. On the other hand, data distinguishing licenses of patents from other licenses are not available; the latter might have increased more rapidly.

Patenting can be an important strategic tool for firms without being either a significant direct stimulus to R&D or a source of technical information on the direction of R&D or other activities of competitors. This appears to be the case in semiconductors and other complex product technologies where it is common for there to be hundreds of patentable elements in one product, with the consequence that no one firm is likely to hold all the rights necessary for a product's commercialization (Cohen et al., 2000; Hall and Ziedonis, 2001). Here mutual dependence on competitors' technologies or mutual vulnerability to other firms' assertions of their patents encourages patenting primarily for the purposes of trading rights, usually by means of cross-licensing arrangements, and avoiding litigation. It is common that in such cross-licensing arrangements one firm pays a royalty to the other firm as a "balancing payment," recognizing the disproportionate strength or impact of the recipient firm relative to the other cross-licensing firm. Nevertheless, the avoidance of litigation is important, since litigation can be especially damaging in an industry where a new product can provoke multiple infringement suits and the capital investment required to produce it is very large.

The pattern of patenting and licensing in semiconductors could represent an active, efficient market in leading-edge technologies or a cost saving relative to litigation. In either case the costs of strategic patenting are not trivial and may redirect resources away from productive research or raise costs to consumers. Moreover, the practice may encourage patent portfolio races among firms trying to gain a negotiating advantage vis-à-vis each other. Participants in the committee's public meetings described this pattern as prevalent in both semiconductors and software, with potential to spread to other sectors on the heels of business method patents. The patent system may also affect the formation of new firms and the innovation associated with entry. Here, too, both theoretical considerations and the extremely limited empirical evidence point in different directions, even in the same industrial context. In semiconductors, for example, the need to have substantial patent assets to trade in order to participate in the pervasive cross-licensing of portfolios probably acts as a barrier to new entrants, although the enormous capital required to establish semiconductor manufacturing capacity is an even more substantial barrier. Yet, in a study commissioned for this project, Ziedonis (2003) suggests that patent protection has been critical to the rapid growth in the number of semiconductor design ("fabless") firms that do no manufacturing. It seems likely that patents have become a more important basis for raising venture capital for biomedical research applications, especially those arising from university activity (Henderson et al., 1999).

Increasing Costs

The direct costs of the patent system are significant, increasing, and in some cases may adversely affect innovation.

The direct costs associated with the acquisition, exercise, and defense of patents are examined further in Chapter 3. Here we simply enumerate some of them to support our proposition that the patent system's evolution merits close attention. First, from the point of view of the inventor or firm applying for a patent, it is estimated that the average corporate U.S. patent prosecution now costs the applicant $10,000-$30,000 in fees. Legal counsel represents the vast majority of that amount, as fees paid to the USPTO are low and have been fairly stable since 1990. The costs at least to large entities of most elements of U.S. patent prosecution have been increasing at an annual rate of 10-17 percent, according to a survey of corporate and private practitioners conducted biannually by the American Intellectual Property Law Association (AIPLA). These figures should be interpreted cautiously, as they represent only two sets of observations over a few years and derive from a nonrandom survey of attorneys.

Corporate managers and attorneys agree that the costs of conflicts over patents have also increased rapidly. The median cost to each party of proceeding through a patent infringement suit to a verdict at trial is at least $500,000 where the stakes are relatively modest. Where more than $25 million is at risk in a patent suit, the median litigation cost is $4 million for each party, according to the AIPLA survey results. Moreover, litigation occupies significant time and attention of business managers and technical personnel, not merely in-house and external counsel, in deciding corporate strategy, participating in depositions, and testifying in court. This process is particularly burdensome for small firms and start-ups with fewer managerial personnel and less access to capital finance (Lerner, 1995). Thus, the direct and opportunity costs of litigation may affect the rate of innovation in ways that are hard to measure or even detect.

A neglected and largely undocumented cost of the patent system is associated with working out licensing arrangements or negotiating royalties or simply fending off threats of infringement. This was highlighted in the Hall and Ziedonis interviews of semiconductor company executives as a significant cost of the current patent-intensive cross-licensing system in that industry despite its relative effectiveness in avoiding the far higher costs of litigation (Hall and Ziedonis, 2001).

3

Seven Criteria for Evaluating the Patent System

PATENTS AND INNOVATION

Ultimately, the test of a patent system is whether it enhances social welfare, not only by encouraging invention and the dissemination of useful technical information but also by providing incentives for investment in the commercialization of new technologies that promote economic growth, create jobs, promote health, and advance other social goals. Assessing the system's overall economic impact is no simple task, perhaps an impossible one. For one thing, the dual functions of patents are in some degree at odds with each other. The exclusivity that a patent confers is undermined by its publication, which may help others circumvent the patent. Furthermore, patents entail a trade-off between the incentives provided for innovation and the costs resulting from a monopoly that may curtail competition and raise consumer prices or hinder further incentive efforts. Both sides of that ledger are exceedingly complex. Innovation in any technology area may benefit from the incentive created by a patent on a new product or process development, but it may suffer if patents discourage the combining and recombining of inventions that would have been made absent the patent or inhibit follow-on discovery. Competition may suffer when an inventor is granted a temporary monopoly right or a combination of patents is used to bar entry or to maintain a cartel in an industry. On the other hand, competition will benefit if this right facilitates investment by new, innovative firms lacking assets other than intellectual property. Patents can also foster the creation of markets for technology, enabling efficiencies in the research and development (R&D) process and promoting the transfer of discoveries from entities skilled at conducting R&D to firms potentially better suited to commercializing and marketing innovations.

We have previously cited evidence that patents function differently in different industrial sectors. There is also a growing body of research on the relationship between patents and innovation across countries and time. Using mainly 19th-century data, Lerner (2002) and Moser (2003) find that instituting a patent system or strengthening an existing patent system does not produce more domestic innovation although the latter does induce inventors from other countries to patent more in the country making the change. It may also induce foreign multinationals to transfer more technology to affiliates in the country (Branstetter et al., 2003). Sakakibara and Branstetter (2001) studied the effects of a statutory change in Japan allowing multiple claims per patent, as has always been the case in the United States. They found that the effective broadening of patent scope had a very small positive effect on R&D activity by Japanese firms. Lanjouw and Cockburn (2000) found some limited evidence for attributing an increase in Indian research addressing developing country needs to patent reforms of the 1980s, which provided increased protection.[1] The effect leveled off, however, in the following decade. Scherer and colleagues (1959) investigated the consequences of Italy's moving from a no-patent to a patent regime in pharmaceuticals; they did not find a significant effect. Using firm-level survey data for Canada, Baldwin and colleagues (2000) found a much stronger relationship running from innovation to patenting than in the reverse direction. Firms that innovate take out patents, but firms and industries that make more intensive use of patents do not tend to produce more innovation. In the United States manufacturing sector, however, in a model that explicitly controls for mutual causation between patenting and R&D, Arora and colleagues (2002) find evidence that patenting is an important stimulus for R&D.

Other positive results are those of Park and Ginarte (1997) using data across 60 countries for the period 1960-1990. They found that the strength of intellectual property (IP) protection (an index of pharmaceutical coverage, participation in international agreements, lack of compulsory licensing, strength of enforcement, and patent duration) was positively associated with R&D investment in the 30 countries with the highest median incomes. Elsewhere, the relationship was positive but not significant. These results, however, are cross-sectional and fail to account for the reverse causality between conducting R&D and having a robust patent system.

The conclusions from this body of empirical research on the effects of patents are several but mostly tentative (Hall, 2003b). In developed countries, at least in manufacturing, patenting stimulates innovative activity broadly, but the stimulus varies among industries. Introducing or strengthening a patent system, however, unambiguously results in an increase in patenting and may encourage the strategic and tactical use of patents with attendant costs and possibly adverse

[1] Although not a level of protection comparable to that in North America, Europe, or Japan.

impacts on innovation and competition. One may legitimately question whether the impact of patenting on innovation and its consequences for social welfare are, on balance, positive outside of the handful of industries, such as pharmaceuticals, biotechnology, medical devices, and specialty chemicals where the benefits are well established, and possibly to a lesser extent, computers and auto parts.

More subtle effects are suggested by recent economic studies and deserve more attention. Patents may enable the creation and affect the organization of knowledge-based industries by allowing trade in knowledge and facilitating the entry of firms with only intangible assets. As this abbreviated literature review suggests, the empirical economic research on the uses and impacts of patenting is more robust than it was nearly 20 years ago when George Priest (1986) complained about the dearth of useful economic evidence on the impact of intellectual property: "Economists know almost nothing about the effect on social welfare of the patent system or . . . other intellectual property." Nevertheless, knowledge is still quite limited and the range of industries examined in any detail is quite narrow.

EVALUATION CRITERIA

In circumstances that at this stage defy a comprehensive evaluation, the committee posits a series of criteria for evaluating the patent system in terms of its impact on innovation rather than addressing its competitive or overall welfare effects. These criteria, although requiring judgments, can in varying degrees be assessed empirically and tracked over time to observe significant changes. In most cases they relate to factors widely thought to be important if not necessary and sufficient conditions for innovation.

First criterion: The patent system should accommodate new technologies. A system granting even temporary monopoly rights to developers of one technology but providing no incentives to developers of other, including substitute, technologies obviously would be hostile to innovation over the long run.

Second criterion: The system should reward only those inventions that meet the statutory tests of novelty and utility, that would not at the time they were made be obvious to people skilled in the respective technologies, and that are adequately disclosed. In the extreme case where an invention is already accessible to the public, or the full scope of what is patented cannot be carried out in practice, there is nothing to be gained and potentially a great deal to be lost by granting a monopoly.

Third criterion: The patent system should serve its second function of disseminating technical information. That means that descriptions of patented inventions should be as complete, clear, and accessible as possible and disclosed in a reasonably timely manner, and there should not be deterrents to consulting the patent or any other technical literature.

Fourth criterion: Administrative and judicial decisions entailed in the patent system should be timely, and the costs associated with them should be reasonable and proportionate. Protracted uncertainty about whether a patent on an application will issue or about whether a patent that is challenged in an infringement dispute will be upheld or found not infringed is not conducive to the investments necessary to innovate. In the same vein, high transaction costs entailed in obtaining or defending a patent are likely to discourage innovation. Such costs tend to escalate the longer the resolution of the issue, whether patentability or infringement, is delayed.

Fifth criterion: In scientific research and in the development of complex or cumulative technologies, where one advance builds upon one or more previous discoveries or inventions and full exploitation of the technology is beyond the capacity of any single entity, reasonably broad access to patented inventions is important. Access depends upon at least three factors: (1) the scope of the patent claims, (2) the availability of licenses on reasonable terms, and (3) the complexity of the patent landscape. Of course, technology must first be created for access to be an issue. Thus, access must be balanced against the incentive to invent and disseminate technology.

Sixth criterion: In an economy where a significant share of its technology-intensive products are bought and sold internationally, the compatibility of national patent systems can be a facilitator of trade and investment and therefore innovation. Indeed, there is an efficiency argument for the integration of the U.S., European, and Japanese patent systems to reduce public and private transaction costs.

Seventh criterion: There should be a level field, with intellectual property rights holders who are similarly situated (e.g., state and private institutions performing research) enjoying the same benefits, while being subject to the same obligations.

Accommodating New Technologies

As the examples of the extensions of patenting in Chapter 2 illustrate, the patent system has proven highly adaptable to changes in technology. This includes not only emergent technologies in advance of or in tandem with their commercial application—for example, biotechnology and nanotechnology—but also technologies that at least in their early stages exhibited rapid progress and substantial commercial success without patents, such as software.

The flexibility of the patent system is a function of at least three features. First, it is a unitary system with few a priori exclusions. Second, the initiative to extend patenting to a new area lies in the first instance with inventors and commercial developers, not with legislators, administrators, or judges. Third, some statutory features of the patent system, as well as administrative and court-interpreted case law, allow for somewhat specialized treatment in some fields of technology.

The Patent Act of 1952 states that

> Whoever invents or discovers any new and useful process, machine, manufacture, or composition of matter, or any new and useful improvement thereof, may obtain a patent therefore, subject to the conditions and requirements of this title.[2]

The most expansive Supreme Court interpretation of this section was in *Diamond v. Chakrabarty,*[3] the case that held a genetically modified microorganism to be patentable subject matter. In the course of its decision the Court stated that

> the Committee Reports accompanying the 1952 Act inform us that Congress intended statutory subject matter to "include anything under the sun that is made by man."

Sometimes these extensions occur readily. The first patent on a flying machine was issued to Orville and Wilbur Wright within 30 months of the flight at Kitty Hawk, North Carolina. In other cases the federal courts have played a prominent role. Particularly when the emergence of a new domain—for example, genetically modified life forms—is obvious and sensitive, the patent office has been hesitant to move in aggressively, and the courts have been asked to recognize patent eligibility. But even in these cases, the lag, if any, can be quite short. The Supreme Court's *Chakrabarty* decision preceded by two years the introduction of the first commercial product, human insulin, made with recombinant DNA techniques.

In other instances the judges have changed their minds over time. With respect to computer software and related inventions, the law changed radically during the latter decades of the 20th century. In the 1970s the Supreme Court held unanimously in *Gottschalk v. Benson*[4] that a computer program was not patentable subject matter. Following two later Supreme Court decisions that suggested a shift in this position,[5] the Court of Appeals for the Federal Circuit ("Federal Circuit") felt comfortable in holding in 1994 that an abstract mathematical algorithm was not patentable, but a computer programmed to run such an algorithm was patentable.[6] This may have been a nearly inevitable development, considering that innovations in the design of the software to run a computer and mechanical devices controlled by internal computer chips seem very close to traditional

---

[2]35 U.S.C. Sec. § 101.

[3]447 U.S. 303 (1980).

[4]409 U.S. 63 (1972).

[5]*Parker v. Flook,* 437 U.S. 584 (1978), and *Diamond v. Diehr,* 450 U.S. 175 (1981).

[6]*In re Alappat,* 33 F.3d 1526, *available at* 1994 U.S. App. LEXIS 21129, 31 U.S.P.Q.2d (BNA) 1545 (Fed. Cir. 1994).

inventions.[7] But the courts have gone even further. The case that has received the most attention is *State Street Bank & Trust v. Signature Financial Group*,[8] which contradicted the prevailing assumption that business methods were not patentable. *State Street* was followed by *AT&T v. Excel Communications, Inc.*,[9] which, in essence, removed the requirement that software could be patented only as embodied in a computer program and therefore effectively permitted patents on algorithms themselves.[10]

Thus, the path toward incorporating new technologies in the patent system is not always rapid and seamless. Even less is it free of controversy. The wisdom of permitting the patenting of inventions involving genetic material, computer software, and especially methods of transacting business, where there is long history of innovations without patent protection, is still very much a matter of debate.[11]

Moreover, the courts have recognized limits to patenting. Historically, patent law has supported the public domain of fundamental scientific research results and other ineligible subject matter not expressed as a product or a method. In its decision in *Chakrabarty* the Supreme Court qualified its "anything under the sun by the hand of man" dictum as follows:

> This is not to suggest that [Section] 101 has no limits or that it embraces every discovery. The laws of nature, physical phenomena, and abstract ideas have been held not patentable. Thus, a new mineral discovered in the earth or a new plant found in the wild is not patentable subject matter. Likewise, Einstein could not patent his celebrated law that $E=mc^2$; nor could Newton have patented the law of gravity. Such discoveries are manifestations of nature, free to all men and reserved exclusively to none.

[7]Indeed, although the European Patent Convention explicitly excludes from patentability "programs for computers as such" (Art. 52(2) and 52(3)), the European Technical Board of Appeals has found it very difficult to keep the exception narrow and has upheld patents to several computer program innovations. For example, International Business Machines, Case No. T0935/97 (Feb. 4, 1999); and International Business Machines, Case No.T1173/97 (July 1, 1998). The European Parliament is currently considering a directive that directly embraces software patents.

[8]*State St. Bank & Trust Co. v. Signature Fin. Group*, 149 F.3d 1368, *available at* 1998 U.S. App. LEXIS 16869, 47 U.S.P.Q,2d (BNA) 1596 (Fed. Cir. 1998).

[9]*AT&T Corp. v. Excel Communications,, Inc.*, 172 F.3d 1352, *available at* 1999 U.S. App. LEXIS 7221, 50 U.S.P.Q.2d (BNA) 1447 (Fed. Cir. 1999).

[10]On remand, the patent involved in this case was held invalid for anticipation and obviousness. *AT&T Corp. v. Excel Communications, Inc., available at* 1999 U.S. Dist. LEXIS 17871, 52 U.S.P.Q.2d (BNA) 1865 (D. Del. Oct. 25, 1999).

[11]Some members of the committee embarked on our study with great skepticism about the wisdom of patenting business methods in the absence of a convincing case for their protection and with some interest in a contemporary proposal to limit the term of business method patents to three or five years. A few members remain convinced that patents are not the most appropriate form of protection for software inventions. Nevertheless, we soon agreed to focus our efforts on means of ensuring better quality business method and software patents rather than on creating exceptions to the general system. The impact of business method patents merits rigorous study after longer experience.

The recent extension of patenting has led to the granting of quite abstract patents, some of them representing intersections of biotechnology, software, and business methods. Examples include the use of a specific genetic characteristic to infer a specific phenotypic characteristic,[12] a technique of statistical analysis on arrays[13] and databases,[14] and the use of specific protein coordinates in a computer program to search for protein complexes.[15] It is of concern to some members of this committee but not clear to a majority that the line between practical invention and pure information is being breached. If it is being crossed in a few cases it is not clear that they represent precedents that the USPTO is continuing to follow, or if the patents were challenged, how the courts would construe these claims or whether the claims so construed would be held valid.[16] That there is disagreement should not be surprising given that the line between ideas and inventions is indistinct.

Notwithstanding its unitary character, the U.S. patent system is differentiated in transparent and subtle ways that accommodate differences in technologies or that affect technologies differently. An example of the former is the requirement for patent holders to pay maintenance fees periodically to take advantage of the full statutory patent term. As we discussed in Chapter 2, that means that many patents are allowed to lapse if the cost of keeping them in force exceeds their value. That is much more frequently the case in information technology, where the product cycle is as short as a few months, than in pharmaceuticals, where the returns to patents are concentrated in the last few years of their terms because the early years are consumed with clinical testing and achieving regulatory approval. The patent prosecution process also varies in duration and other characteristics from one major technology class to another (Allison and Lemley, 1998).

Less obvious but important, the patentability rules applied to different technologies show some divergence. According to legal scholars Dan Burk and Mark Lemley (2003a), the ability to calibrate the patent system to industries and technologies derives from a large kit of policy levers available to the USPTO and the courts. These include or could include all of the following rules and patent doctrines—the rule against patenting abstract ideas, the standard of utility, the exception for experimental use, the test for obviousness of the "person having

---

[12]U.S. Patent No. 5,998,145.

[13]U.S. Patent No. 6,647,341.

[14]U.S. Patent No. 6,023,659.

[15]U.S. Patent No. 6,252,620.

[16]For example, there are at least two patents with at least one claim to computer-readable material encoded with protein structure coordinates (U.S. Patent No. 6,546,074 and U.S. Patent No. 6,389,378) that could be at odds with USPTO examination guidelines. See I. Shimbo et al. (2004), which reports the results of a 2002 trilateral (USPTO, JPO, and EPO) review concluding that "information" such as protein three-dimensional structural coordinates is not patent-eligible subject matter in any of the three jurisdictions.

ordinary skill in the art," so-called secondary considerations of non-obviousness (for example, commercial success, long-felt need), the written description requirement, the doctrine of equivalents, the principle of pioneering patents, the presumption of validity, patent misuse, and injunctive relief.[17] Often their application, not just the technology, is controversial, but they give the patent system a flexibility that would be lacking if it were necessary to amend the patent law every time a new technology presented itself.

Ensuring High-Quality Patents

In 1790 when Congress enacted the first patent statute it stipulated two substantive requirements—novelty and utility—for an invention or discovery to qualify for a patent. From the outset it was recognized that patents ought not to be granted for any trivial advance in an art, that some more substantial improvement should be shown. In 1851 the Supreme Court distinguished the "work of the skillful mechanic," not justifying protection, from the "work of the inventor"; but for a century, courts struggled without statutory guidance to define an "invention." Finally, in 1952, Congress adopted an alternative formulation, excluding from patentable subject matter what "would have been obvious at the time the invention was made to a person having ordinary skill in the art." Thus, the third substantive requirement for patentability became known as the "non-obviousness" standard.[18]

The importance of these three conditions in the abstract is uncontested. Patents on known or only trivially modified inventions would confer potential market power to restrict access and raise prices and enable the patent holder to use litigation as a competitive weapon without providing incentives for making genuine advances or disclosing such advances to the public. They offer no public benefit in exchange for the benefit given to the patentee. Granting patents for inventions that are not new or useful or that are obvious unjustly rewards the patent holder at the expense of consumer welfare (Levin and Levin, 2003).

A second theoretical argument against poor patents is that because of doubts about their validity they are likely to encourage more infringement and more litigation, raising the transaction costs of the system and discouraging some investment (Merges, 1999; Meurer, 1989). Poor patents may induce investment

---

[17]Burk and Lemley go on to argue that some of the ways that the courts have applied the legal standards of obviousness, enablement, and written description are misguided—for example, producing more and narrower biotechnology patents and fewer broader software patents whereas innovation policy considerations suggest that the results should be the reverse. Burk and Lemley have been criticized (Wagner, 2003) for not distinguishing between their insightful descriptive "micro-exceptionalism" and their prescriptive "macro-exceptionalism," calling on the courts to play a policy role for which, arguably, they are not suited.

[18]The corresponding European requirement is that a patent application show an "inventive step."

in product development that is abandoned later when the patents are invalidated. Hunt (1999) and O'Donoghue and colleagues (1998) conclude from slightly different models of innovation that raising or lowering the standards of patenting could affect the character of R&D. If the standard is high, firms may be more likely to pursue larger innovations.

Over the past decade the quality of issued patents has come under sharp attack.[19] The conjecture that patent quality is declining or is simply too low has been characterized in two ways. First, some legal scholars have suggested that the standards of patentability—especially the non-obviousness standard—have been relaxed as a result of court decisions (Barton, 2000; Dreyfuss, 1989; Lunney, 2001). Other observers have suggested that the USPTO too frequently—or more frequently than in the past—issues patents for inventions that do not conform to generally accepted standards for patentability, especially in technology areas that are newly patentable, notably genomics, software, and business methods (Barton, 2000; Hall, 2003b).[20] This alleged decline in USPTO performance is variously attributed to the quantity and quality of relevant resources, examiner qualifications, experience and incentives, the time devoted to searching and evaluating each application, and the information available to examiners (for example, access to automated data bases incorporating prior art). Although logically distinct, the notion that standards for patentability are slipping and the notion that USPTO examiners are failing to apply the legal standards appropriately are obviously difficult to distinguish in practice (Cohen et al., 2002).

There is no lack of examples of issued patents that appear dubious on their face. One such list (Hall, 2003b) includes a patent on a computer algorithm for searching a mathematical textbook table to determine the sine or cosine of an angle,[21] a patent for cutting or styling hair using scissors or combs in both hands,[22] a patent on storing music on a server and letting users access it by clicking on a list of the music available,[23] and a patent on initiating forward motion on a child's swing by pulling on the ropes and swinging sideways (the last subsequently ordered to be re-examined by the director of the USPTO).[24] Whether these are products of the office's interpretation of court decisions or of internally generated guidance given to examiners or of less than thorough examination of applications

[19]The complaint is not new, but previously it was associated with periods, such the 1970s, of generally low regard for the patent system, high rates of invalidity determinations by the courts, and low patenting activity.

[20]Such criticisms have been leveled by the Supreme Court. For example in *Graham v. John Deere Co.*, 383 U.S. 1 at 18 (1966), the Court referred to "a notorious difference between the standards applied by the Patent Office and by the courts."

[21]U.S. Patent No. 5,937,468.

[22]U.S. Patent No. 6,257,248.

[23]U.S. Patent No. 5,963,916.

[24]U.S. Patent No. 6,368,227.

or, indeed, whether some of them could withstand challenges in the courts is an open question. Further, whether the examples are aberrant or typical or, for that matter, increasing or declining in frequency is impossible to determine on the basis of a few handpicked examples of apparently bad results. But a nontrivial number of errors in judgment are inevitable in a system whose output by 3,000 individual examiners is 167,000 patents annually.

In the late 1990s the U.S. Department of Commerce inspector general's (IG) office investigated the growing backlog of applications awaiting decisions before the USPTO Board of Patent Appeals and Interferences (U.S. Department of Commerce, 1998). The IG reported that board personnel attributed declining production to the poor quality of cases being appealed.

> Board personnel whom we interviewed stated that cases they receive from the examining corps often contain administrative errors, inadequate support for the examiner's final rejection, and other unanswered questions or omitted information about the patent's claim that should have been addressed. As a result, APJs [administrative patent judges] are spending time searching prior art (technical literature including prior-issued patents and foreign patents, related documents, and non-patent literature such as journal articles and abstracts), a task which is normally examiner responsibility. Board workload data supports their assertions. Reversals of examiner decisions and remands for additional examiner review combined for 41 percent of the board's total disposals in FY 1994, but 54 percent in FY 1997.[25] Furthermore, rejections due to examiners having overlooked prior art have averaged 12 percent of the board's decisions over the same period. In effect, overall production is cut because APJs are spending more time processing appeals in order to make these determinations.

Nevertheless, the claim that quality has deteriorated in a broad and systematic way has not been empirically tested. Three seemingly direct measures of quality are (1) the ratio of invalid to valid patent determinations in infringement lawsuits, (2) the error rate in USPTO quality assurance reviews of allowed patent applications, and (3) the rate of claim cancellation or amendment or outright patent revocation in re-examination proceedings in the USPTO.[26] These indicators show mixed results. The rate of invalidity findings in district (trial) court judgments has declined over time. P. J. Federico (1956), using data for 1925-1954, and Gloria Koenig (1980), using data for 1953-1978, found that before 1982 district courts and circuit courts upheld only about one-third of the patents litigated. At the appeals level the rate increased to about 55 percent with the advent of the Federal Circuit (Dunner et al., 1995), as did the validity rates in the district courts as a whole (Lemley, 2002, using data from 1994; and Allison and

[25]The USPTO Annual Report stated the combined reversal/remand rate was slightly less in FY 1997—51 percent.

[26]See Appendix A for a description of the re-examination procedure.

Lemley, 1998, using data from 1989 through 1996).[27] Although it may seem surprising that the probability that a patent will hold up under challenge is just over 50 percent, it should not be unexpected. Both parties exercise enormous care in deciding whether to run the risk of litigating a patent dispute rather than abandoning or settling it, the much more frequently exercised options. In most cases, not only is the commercial value high but also the validity issues are finely balanced. Consequently, one should be very cautious in interpreting the results of courts' validity decisions.

The error rate reported in USPTO quality assessment audits has fluctuated between 3.6 and 7 percent since 1980. There was a slight upward trend through the 1990s until 2000, but it has declined in recent years to around 4 percent. Only about 10 percent of patents subject to re-examination in the United States are completely revoked, although nearly two-thirds undergo some adjustment to their claims, often because the patent holders themselves sought re-examination to modify their claims in light of newly discovered prior art.

All three indicators suffer from serious deficiencies, however. In addition to selection effects, the numbers of patents subject to any of these procedures are extremely small. The litigation rate of issued patents is just over 1 percent (Lanjouw and Schankermann, 2003); re-examined patents represent about 0.3 percent of the total (Graham et al., 2003); and about 2 to 3 percent of a year's patents are reviewed by the USPTO for quality control purposes.

Ostensibly, the USPTO's audits come closest to producing a measure of quality and therefore deserve closer examination. The patents reviewed are not randomly chosen to assess overall system performance nor is the selection weighted toward technologies in which examination quality may be problematic. Currently, the protocol is designed to take a specific number of applications from each examiner depending upon examiner experience level and certification status.[28] Because of the small percentage of allowed applications that are reviewed, the error rates are statistically significant only at the level of the seven technology centers, not the art units.[29]

---

[27]The Federal Circuit is much more likely to affirm a district court's finding of validity than invalidity. This is a reversal of the previous relationship between the district courts and circuit courts of appeals.

[28]Under the USPTO's 21st Century Strategic Plan, experienced examiners are recertified for competency every three years. Recertification is based in part on an expanded review of the examiner's recent work by both the Technology Center Management and the Office of Patent Quality Assurance.

[29]Curiously, error rates have tended to be higher in technologies where examination is most straightforward and least complicated. This may be because examiners in these technologies have less time allotted for examining each application or because reviewers find it easier to review and understand these applications and therefore more easily recognize errors. With respect to this latter possibility it should be noted that reviewers are drawn from the technology class they review (that is, chemical, electrical, mechanical), but they are required to review applications covering much broader subject matter than examiners are required to examine.

In any case, the history of the USPTO's quality review function does not inspire confidence that its results are meaningful and consistent over time. Created in 1974 in response to earlier criticisms of patent quality, the Office of Patent Quality Review was twice reviewed harshly by the inspector general of the Department of Commerce. In 1990 the IG faulted the USPTO for failing to reduce error rates by using data from the quality review process. Although rates did decline from Fiscal Year (FY) 1991 to FY 1996, the quality review staff was reduced by one-half, as was the sample rate, from nearly 4 percent of patents to just over 2 percent, too low to provide valid results for any of the art units, according to the IG. Meanwhile, the USPTO management proposed to eliminate the quality review auditing of issued patents and "in process" reviewing as well as to substitute a survey of "customer" (that is, patentees') satisfaction. A second IG report (1997) criticized both the deterioration in the auditing function and the unreliability of the proposed alternative. The USPTO agreed to reestablish a "strong, independent" Office of Patent Quality Review. As a result of the 21st Century Strategic Plan, quality assurance specialists with principal responsibility for the auditing function have been deployed to the technology centers although they report to the Office of Patent Quality Assurance, reporting in turn to the deputy commissioner for patent operations. Previously, the Office of Patent Quality Review was entirely independent of the patent administration. Whether this shift changes auditors' incentives remains to be seen. It may facilitate communication with examiners and managers.

Another way to test empirically whether there has been a change in patent quality would be to "peer review" a representative sample of patents in different technical areas from different time periods. A group of experts independent of the USPTO could rate the patents on novelty, utility, obviousness, and quality of the description. That has not been done because it is a substantial undertaking but one worth consideration.

What about indirect measures of quality? In research supported by this project Allison and Tiller (2003) examine prior art references in Internet business method patents, one of the categories of patents whose quality is most suspect. They compare the number of references (that is, backward citations) in their sample to those found in a random sample of all other patents. They find that the business method patents contained substantially more total references and patent and nonpatent references than the patents in the general sample. This finding runs counter to the widely held assumption that the USPTO has consistently overlooked nonpatent prior art in the examination of business method applications. Nevertheless, Allison and Tiller's data cannot answer several intriguing questions. For example, is the body of nonpatented prior art in the area of business methods so large or diverse that examiners are still missing a good share of it? Does the examination process overlook some business methods that are in common use but not documented in written sources?

There are several reasons to suspect that more issued patents are deviating from previous or at least desirable standards of utility, novelty, and especially non-obviousness and that this problem is more pronounced in fast-moving areas of technology newly subject to patenting than in established, less rapidly changing fields.

Workload Pressures on the USPTO

One reason for suspecting that quality has suffered is that even before taking examiner qualifications and experience into account, the number of patent examiners in recent years has not kept pace with the increase in workload represented not only by the enormous growth in the number of applications (doubling in 10 years, between 1991 and 2001) but also by the growing complexity of applications as represented by the growth in the number of claims and prior art citations per application (Allison and Lemley, 2002).[30] The number of examiners per 1,000 patent applications is down about 20 percent over the last four or five years (see Figure 3-1) in part because of congressional reluctance to increase personnel. At the same time, the Congress has for several years appropriated a portion of the fees collected by the USPTO to other governmental activities.

It may be that examiner productivity has improved somewhat with access to scientific and technical literature databases capable of automated search for prior art, but a potentially more important source of productivity gains—automated filing and processing of applications—is only now being implemented on a large scale.

The principal result of holding employment growth well below the growth in applications has been longer pendency, rising from an average of 18.3 months in 1990 to 24 months in 2002. The average time an examiner spends on an application has remained largely unchanged,[31] meaning that the volume of work may have been accommodated without serious detriment to examination thoroughness, but there has been no apparent adjustment across all technologies to compensate for the greater complexity of the average application.

---

[30]Is the growing complexity of applications itself a sign of deteriorating or increased quality or neither? It may be argued that longer applications are easier to draft than shorter ones and allow concealment of the important features of an invention. On the other hand, longer filings may reflect more disclosure and claim refinement. The answer has implications for whether the USPTO should accommodate or even encourage the trend (for example, by allowing more examination time) or penalize it as first versions of the 21st Century Strategic Plan proposed to do by charging higher fees. The issue deserves further study.

[31]Estimates of the average examination time per patent range from 15-17 hours on the low side (testimony of H. Manbeck, former commissioner of the USPTO [*Boehringer Ingelheim Vetmedica, Inc. v. Schering-Plough Corp.*, 68 F. Supp. 2d 508 at 525, *available at* 1999 U.S. Dist. LEXIS 16989 (D.N.J. 1999)]) to 25-30 hours on the high side (Barton, 2000).

FIGURE 3-1 USPTO examiner workforce. SOURCE: USPTO.

The number and time allocation of examiners says nothing about their training, qualifications, experience, length of tenure, compensation, and performance evaluation criteria. It may well be that thorough examination of these organizational and workforce characteristics—which we were unable to undertake—would reveal other reasons to be concerned about patent quality as well as important ways to improve it.[32]

Patent Approval Rates

A second reason for concern about changes in quality is that patent approval rates may be significantly higher than officially reported by the USPTO. For a number of years the USPTO has reported that approximately two-thirds of patent applications result in patent grants. In a recent study Quillen and Webster (2001) argued that calculations of allowance rates from USPTO reported numbers of applications filed, abandonments, and total allowances or issued patents have led to a consistent underestimate of actual allowance rates because the calculations did not take into account the effect of U.S. continuation practice. By statute the

[32]USPTO management has been attentive to some of these variables, seeking authority for higher pay and instituting an examiner recertification requirement.

United States allows applicants to refile applications to obtain continued examination of the invention claimed in the original application (see Appendix A). Since more than one application claiming a specific invention may be filed before a patent is granted, calculations that do not correct for continuation applications underestimate the allowance rate. Quillen and Webster concluded that once continuation, continuation-in-part, and divisional applications[33] are accounted for as renewed attempts to protect the subject matter of their applications, the USPTO eventually issued patents on between 85 percent and 97 percent of applications filed between 1993 and 1998—20 to 30 percent higher than official estimates.

Quillen and Webster noted the possibility that more than one patent could issue from a single disclosure, but because they did not have the data to correct for such occurrences, they based their calculations on the assumption that "parent" patent applications are abandoned when a continuation application is filed. In a follow-up paper Quillen, Webster, and Eichmann (2002) attempted to account for applications that give rise to more than one patent by using a random sample of 1,000 patents developed by John Allison and Mark Lemley to determine the percent of those that were granted on continuations whose parents were also issued as patents. When they incorporated a correction for all continuing applications, including divisionals and continuations-in-part, they calculated an allowance rate of 83 percent.

Last year Robert Clarke (2003) of the USPTO published a review of Quillen and Webster's original findings along with his own analysis of USPTO allowance rates. Like Quillen and Webster, Clarke subtracted continuation applications from the total applications to derive the number of "original applications," but he also subtracted from the total pool of patents issued during the relevant time period all patents issued from applications with an ancestor that had also issued into a patent. Clarke's analysis benefited from additional USPTO data not available to Quillen and Webster. Clarke concluded that the likelihood of a U.S. patent grant from an original application for applications filed during the five-year period from 1994 through 1998 was slightly less than 75 percent. Clarke attempted to validate his calculations by also determining the percentage of applications that go abandoned without being refiled. The percentage of applications that do not issue as patents or give rise to further continuing applications was found to be slightly greater than 25 percent, complementing the percentage of allowed applications determined by counting only those patents that issued from applications that were not continuations of applications that also issued into patents.

The methods of Clarke and those of Quillen and colleagues necessarily rely on certain assumptions, mainly to account for their inability to follow individual

[33]See Appendix A for definitions of these terms.

applications and application families from original filing to final disposition of all members. For this reason arriving at a consensus on a precise patent approval rate may be elusive. Nevertheless, we can infer from these efforts that the ability to file continuation applications with the USPTO gives applicants a higher probability of obtaining patents on some version of their inventions.

Acceptance rates by themselves ignore how patent claims are modified, nearly always by narrowing their scope, in the course of the examination, surely a key determinant of quality. Moreover, rigor of examination is only one of several factors that may affect allowance rates. The fact is that the examination procedure, allowing an applicant multiple attempts to persuade a critical examiner to approve a patent (see Appendix A), is designed to yield a high "success" rate, at least for persistent applicants.[34] The predictability of the standards in a particular technology and the perceived economic value of the patent are some of the factors that affect motivation to pay the costs associated with that iterative process.

The committee believes that high acceptance rates, especially if increasing over time relative to comparable rates in other industrialized countries would be reason to look more closely at examination quality. Under either Quillen and Webster's or Clarke's assumptions the USPTO patent approval rate gradually increased in the early 1990s and then declined after 1998 (see Figure 3-2). The European Patent Office (EPO) and Japanese Patent Office (JPO) approval rates peaked at approximately the same time but then declined more rapidly, so that in 2000 the USPTO rate was higher although by a substantial margin only under Quillen and Webster's assumptions. On the other hand, the Organisation for Economic Co-operation and Development (OECD, 2003) estimates that the USPTO grant rate for U.S. priority applications with at least one subsequent EPO application was consistently higher than the EPO grant rate for U.S. priority applications throughout the 1980s and early 1990s—80 to 90 percent versus 50 to 60 percent.[35]

These analyses have given the USPTO tools to make more realistic comparisons than the officially reported statistics. These tools should be applied to determine acceptable rates in different technology classes, especially ones newly subject to patenting. If increases in allowance rates are found, other potential causes need to be considered, of course. For example, it is possible that the higher cost of obtaining patents has caused firms to be more rigorous in screening inventions for which they file applications, or that greater predictability in the applications

---

[34]As Lemley and Moore (2004) observe, "One of the oddest things about the U.S. patent system is that it is impossible for the U.S. Patent and Trademark Office to ever finally reject a patent application. While patent examiners can refuse to allow an applicant's claim to ownership of a particular invention, and can even issue what are misleadingly called 'Final Rejections,' the patent applicant always gets another chance to persuade the patent examiner to change her mind."

[35]Differences in patent office practices—for example, Japan's narrower claiming—may affect these comparisons.

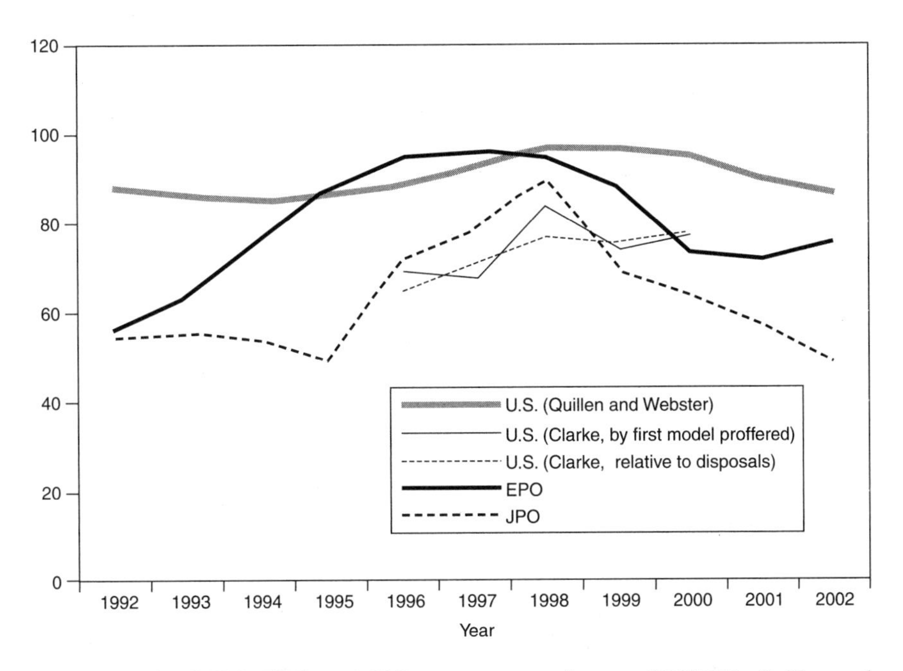

FIGURE 3-2 USPTO, EPO, and JPO patent approval rates. SOURCE: Quillen and Webster (2001) and Clarke (2003).

of patentability criteria by the USPTO means that firms are better at weeding out inventions that will not result in granted patents.[36]

Changes in Treatment of Genomic and Business Method Inventions

Partially in response to criticisms of the standards being applied to business method and genomic patent applications, the USPTO conducted a broad review of those categories and instituted significant changes in procedures and standards. In March 2000 the Patent Office announced the "Business Methods Patent Initiative" focused on Class 705 ("Data Processing: Financial, business practice, management, or cost/price determination"), encompassing the bulk of the business method applications filed in the wake of the *State Street Bank* decision and many of the well-known Internet patents including Amazon's "one-click" shopping

[36]For example, the OECD study results could be explained by a more rigorous screening by applicants of candidates that justify the high cost of foreign filing. U.S. firms may be better able to predict outcomes in the USPTO but less able to predict outcomes in foreign patent systems. Most foreign systems lack a grace period so that applicants face more prior art than in the United States.

method[37] and Open Market's "on-line shopping cart."[38] The initiative consisted of four steps: (1) improved technical training of Class 705 examiners, (2) revised examination guidelines, (3) mandatory search of specified sources of prior art, and (4) a new "second review" of all allowed applications to ensure compliance with the search guidelines and the appropriateness of allowed claims.

In the following January 2001 the USPTO responded to similar criticisms of the patents being allowed on human genetic sequences by releasing new guidelines clarifying the written description and utility requirements. The guidelines are written to be generic to all technologies, but most affected are claims involving DNA and proteins, and most of the training examples are in biotechnology. The written description guidelines were intended to bring USPTO practice into line with the Federal Circuit's decision in *Regents of the University of California v. Eli Lilly and Co.*,[39] stating that simply describing a method for isolating a gene or other sequence of DNA is insufficient to show possession, and the complete sequence or other identifying features must be disclosed. The utility guidelines declared that the claimed utility of the invention must be "specific, substantial, and credible" and extend beyond merely describing its biological activity. The guidelines were widely interpreted as raising the bar to patents on genomic inventions.

The new policies reflected a recognition by USPTO management that standards needed to be tightened, at least in two technologies attracting large investments and a great deal of publicity and exhibiting a controversial surge in patenting activity. The question of what practical effect the measures had on examiners' behavior and USPTO output is difficult to answer. It is complicated by the lag between application filings and patent grants, the downturn in the economy and in technology investments that occurred in 2000, and other nearly simultaneous developments affecting patenting activity in these fields. For example, at the same time that DNA patent applications were accelerating, the international Human Genome Project was rapidly depositing human DNA sequence data in the public domain, where it became prior art. A "working draft" of the genome was published in February 2001.

Class 705 patent grants peaked in the last quarter of 1999 and fell sharply in the first quarter of 2000, coincident with the institution of the second review and other measures.[40] The decline continued throughout 2000 before leveling off.

---

[37]U.S. Patent No. 5,960,411.

[38]U.S. Patent No. 5,715,314.

[39]*Regents of the Univ. of Cal. v. Eli Lilly & Co.*, 119 F.3d 1559, *available at* 1997 U.S. App. LEXIS 18221, 43 U.S.P.Q.2d (BNA) 1398 (Fed. Cir. 1997).

[40]Of course, it is misleading to suggest that policy changes occur at a single point in time. They are preceded by much discussion, including discussion internal to the USPTO; and the public announcement is followed by a period of implementation. Thus, some examiners may have anticipated the policy shift on business method patents; others may not have complied with it immediately. The same considerations apply to the tightening of requirements for DNA patents.

Classification 536/23.1 ["DNA or RNA fragments or modified forms thereof (e.g., genes, etc.)"] showed a more modest decline in patent grants beginning in 2001, coincident with the new examination guidelines (see Figure 3-3).

For certain companies known to be patenting large numbers of DNA sequences, however, the decline began at least one year earlier (see Figure 3-4). A full assessment of the effect of the written description and utility guidelines on patents on nucleic acids would require an analysis of the scope of issued claims and the types of nucleic acids claimed (e.g., full-length coding sequences, ESTs, antisense fragments with therapeutic potential) in addition to the numerical

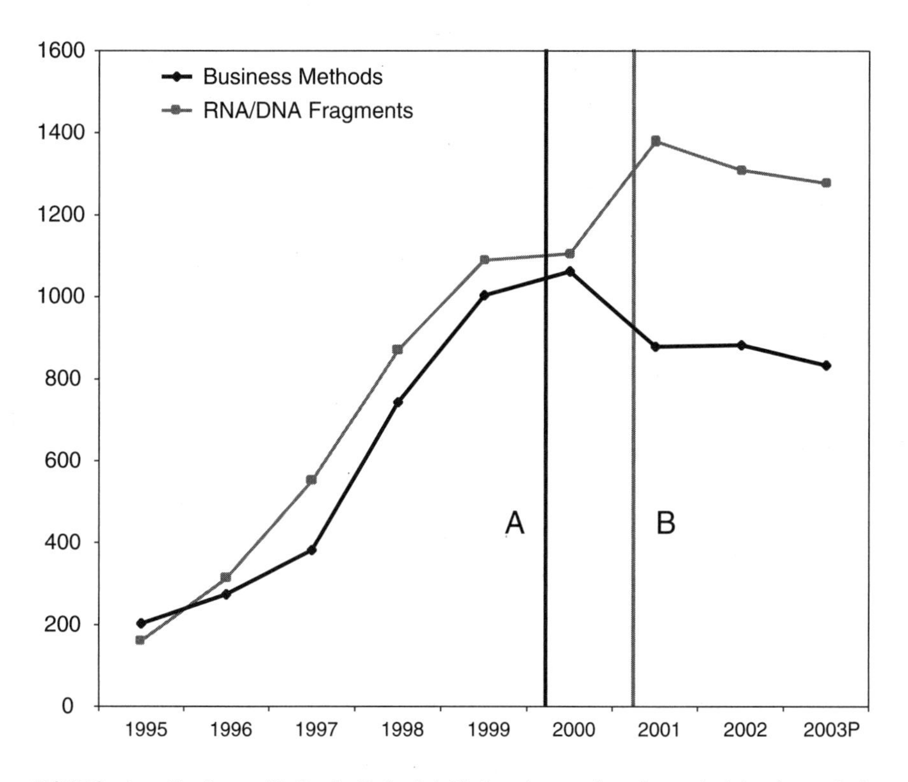

NOTES: A = Business Methods Patent Initiative (second review, etc.) implemented. B = Utility and written description guidelines implemented.

FIGURE 3-3 Business method (USPTO Classification 705) and DNA/RNA fragment (USPTO Classification 536/23.1) patent grants.[a] SOURCE: USPTO.

[a]If data on all patents in the DNA Patent Database, Georgetown University, are used, the same break in the upward trend of patent grants occurs early in 2001. The database is a product of screening several relevant patent classes in USPTO data to yield a set of DNA-related patents.

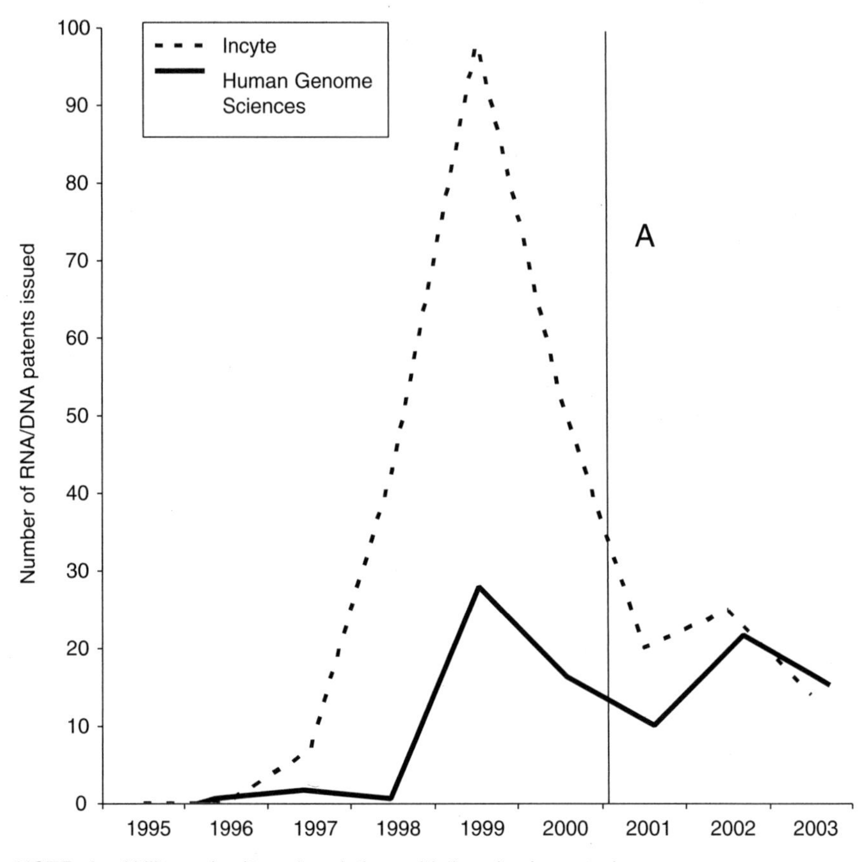

FIGURE 3-4 DNA/RNA fragment patent grants to genomic companies (USPTO Classification 536/23.1). SOURCE: DNA Patent Database, Georgetown University. The database is a product of screening several relevant patent classes in USPTO data to yield a set of DNA-related patents.

analysis shown in Figure 3-3. With respect to DNA patents other factors to be weighed in interpreting patent grants over time is the finite nature of the human genome (an estimated 30,000 genes in all) and the USPTO's "restriction" practice of forcing patent applicants to separate DNA sequences into different applications. The latter is controversial in the biotechnology industry because it raises the cost of obtaining patents, but by simplifying the task of examiners it is more likely to enhance the quality of the results than to degrade it.

It is clear that in neither case did the high-tech economic collapse play a significant role in the slowdown in patent approvals, at least initially. That is because the patents in both categories that were issued in 2000 and 2001 derived from applications filed at least two years earlier, at the height of the boom. It is nevertheless conceivable that the principal effect of the new policies in both cases was to make long pendencies even longer. By the end of 2002 applications in both Class 705 and Class 536/23.1 were taking more than three years to yield patents (see Figure 3-5).

Application of the Non-Obviousness Standard

A fourth reason to be concerned about patent quality is that there may have been some dilution of the non-obviousness standard as a result of court decisions and their incorporation in the examination guidance compiled in the USPTO's Manual of Patent Examining Procedure (MPEP).

Added to the patent statutes in 1952, the standard is stated as follows:[41]

> A patent may not be obtained though the invention is not identically disclosed or described as set forth in section 102 of the title, if the difference between the subject matter sought to be patented and the prior art such that the subject matter as a whole would have been obvious at the time the invention was made to a person having ordinary skill in the art to which the said subject matter pertains.

The enactment of Section 103 was in part a reaction to a line of Supreme Court cases in which patents were held to be invalid because they lacked "invention." In one case Justice William O. Douglas maintained that for an invention to be patentable it "must reveal the flash of creative genius."[42] Justice Robert H. Jackson in a dissenting opinion complained about this trend in decisions by observing that "the only patent that is valid is one which this Court has not been able to get its hands on."[43]

Although it may have been adopted to moderate the antipatent tendency of the Court, Section 103 establishes a level of development beyond not only the documented prior art but also the practice of people of ordinary skill in that art that must be accomplished before a patent can issue. Merges and Duffy (2002) characterize it as the "nontriviality" requirement of the patent law. The Supreme Court did not address the question of how to interpret Section 103 until 14 years after its enactment, when it decided three patent cases frequently referred to as the "*Graham* trilogy."[44] The Court confirmed the abandonment of the notion of

---

[41]35 U.S.C. § 103(a).

[42]*Cuno Engineering Corp. v. Automatic Devices Corp.*, 314 U.S. 84 , 62 S. Ct. 37, *available at* 1941 U.S. LEXIS 1250, 1942 Dec. Comm'r Pat. 723, 51 U.S.P.Q. (BNA) 272 (1941).

[43]*Jungersen v. Ostby & Barton Co.*, 335 U.S. 560, 571 (1949) (dissenting opinion).

[44]*Graham v. John Deere Co.*; *Calmar, Inc. v. Cook Chemical Co.*, 383 U.S. 1 (1966); and *United States v. Adams*, 383 U.S. 39 (1966).

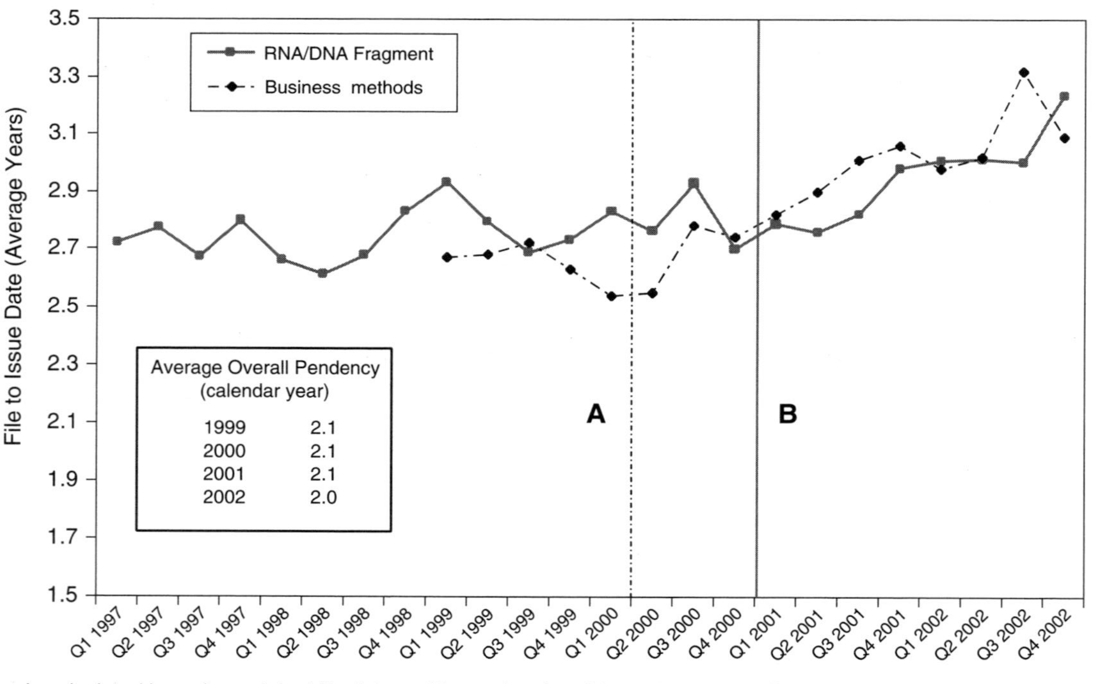

NOTE: Pendency is calculated based on original file date and issue date for all issued patents in Class 536/23.1 and Class 705. Overall pendency is calculated by the USPTO and also includes an estimate of the time from filing date to abandonment of the application.
A = Business Methods Patent Initiative (e.g., second review) implemented.
B = Utility and written description guidelines implemented.

FIGURE 3-5 Class 705 and Class 536/23.1 pendency by quarter.

"invention" as leading to conceptual confusion but said that Section 103 did not and constitutionally could not lower the patentability standard. Indeed, in ruling invalid two of the three patents at issue in the cases, the Court provided the following guidance for evaluating a patent for obviousness:

> [T]he scope and content of the prior art are to be determined, differences between the prior art and the claims at issue are to be ascertained; and the level of ordinary skill in the pertinent art resolved. Against this background, the obviousness or non-obviousness of the subject matter is determined. Such secondary considerations as commercial success, long felt but unsolved needs, failures of others, etc. might be utilized to give light to the circumstances surrounding the origin of the subject matter sought to be patented.[45]

How this blueprint is applied can affect the incentives of both initial and follow-on innovators and the benefits and costs of the patent system. If the required step is too small, the pioneering inventor must share royalties with improvers who might otherwise be excluded. For subsequent inventors the step required affects the choice between seeking ambitious or marginal improvements. Moreover, if the required step is very small, there may result a proliferation of patents that entail costly licensing negotiations and payments and limit firms' future freedom of action. Patents on trivial inventions may confer or help to sustain significant market power. At the same time, an overly restrictive non-obviousness standard could discourage investment and delay new entrants to a market.

Although not in complete agreement about which aspects of which decisions are responsible, a number of legal scholars view the evolution of the law over the last generation as reducing the size of the step required for patentability under the non-obviousness standard and as allowing the issuance of patents on obvious inventions (Barton, 2003; Desmond, 1993; Kastriner, 1991; Lunney, 2001; Merges, 1999; and Vermont, 2001). Since *Graham* there have been four cases in which the Supreme Court has considered obviousness decisions by the circuit courts of appeal. In all four cases the Court found obvious patents that the lower court had held valid, although one of the cases was decided on procedural grounds (Barton, 2003).[46] The Court, however, has not revisited obviousness for nearly two decades.

A 1995 study of Federal Circuit decisions rendered on cases originating in lower courts shows that the court upheld 86.8 percent of decisions holding valid patents faced with non-obviousness challenges, but upheld only 59.9 percent of

---

[45] 383 U.S. 17-18.

[46] *Dennison Mfg. Co. v. Panduit Corp.*, 475 U.S. 809 (1986) (per curiam) (determination of procedural issue); *Sakraida v. Ag Pro Inc.*, 425 U.S. 273 (1976) (combination patent); *Dann v. Johnston*, 425 U.S. 219 (1976) (equivalency as alternate grounds); *Anderson's Black Rock, Inc. v. Pavement Salvage Co.*, 396 U.S. 57 (1969) (combination).

those decisions holding patents invalid on non-obviousness grounds (Dunner et al., 1995).[47] Both rates are higher than the appeals court's overall rate of affirmance in patent cases, which is around 50 percent. With respect to decisions on appeal from the USPTO, the study shows that the Federal Circuit became slightly stricter with respect to non-obviousness and upheld more USPTO rejections during the late 1980s, but then reversed a greater share of USPTO non-obviousness rejections during the early 1990s (Dunner et al., 1995). A more recent study finds a decline in invalidity decisions based on obviousness by the Federal Circuit (Lunney, 2001).[48] There is also evidence that the Federal Circuit judges appointed more recently are more likely to uphold a patent against a non-obviousness argument.[49] Taking into account trial courts as well as the Federal Circuit, non-obviousness is the leading basis of patent invalidity, providing a basis in 42 percent of invalidity findings, and non-obviousness arguments are accepted 36.3 percent of the time (Allison and Lemley, 1998).

Although the committee considered these analyses, it did not reach a position on their significance with respect to non-obviousness generally. Nevertheless, we are concerned about trends in the application of the obviousness standard to business method and genetic sequence inventions. As the problem in each of these areas is different and the recommended solutions are different, both are addressed in Chapter 4.

Neither USPTO resources in relation to its workload, nor patent approval rates, nor changes in the treatment of genomic and business method inventions and the non-obviousness standard are, separately, conclusive evidence that patent quality is too low or declining. However, together they lead the committee to conclude that there are reasons to be concerned about both the courts' interpretations of the substantive patent standards, particularly non-obviousness, and the USPTO's application of the standards in examination. This may be primarily an issue in emerging technologies, where fairly broad patents may be granted early on, and fewer but narrower patents are granted as the field matures, more prior art becomes available, and examiners become more familiar with it. Does this mean that the system automatically adjusts without any need for examiners to be more cautious in issuing patents and the courts more cautious in ruling on validity in a

---

[47]This calculation uses the numbers in Table 2 on page 163, combining appeals from the district courts, the International Trade Commission, and the Court of Claims. Vacated decisions are treated as equivalent to a reversal.

[48]Courts of appeal do not control the issues parties present and when multiple grounds of invalidity are presented will often, once one ground of invalidity is found, decline to address the remaining grounds as moot.

[49]In a study of cases between 1989 and 1996, judges appointed before 1982 rejected non-obviousness arguments in 31 out of 61 votes (50.8 percent), while judges appointed later rejected the arguments in 93 out of 140 votes (66.4 percent) (Allison and Lemley, 2000). For these numbers, P2 = 5.655, indicating that the difference is valid at about the 2.5 percent level.

new technology? That is perhaps the pattern, but the cost of waiting for an evolutionary process to run its course may be too high. As the examples of the Internet technology and biotechnology illustrate, because of the efficiency of U.S. capital markets and the growth of early-stage financing, the stakes become very high very early in the development of new commercial technologies.

Disseminating Technical Information

Disclosure is the quid pro quo for patenting, but patents appear to be a relatively minor means of diffusing technological know-how, possibly less important in the United States than in other countries (Cohen et al., 2000). There are a number of reasons for this, some of them either of little concern or unavoidable. In the United States especially, there is an enormous scientific and technical literature, a tradition of personal communication through technical meetings and conferences, a pattern of interaction between the Academy and industry by means of consultancies, liaison programs, and funding arrangements, and in some geographical regions even a culture that encourages informal exchange of proprietary information between employees of competing firms.[50]

While alternative means of technological diffusion or, in economists' terms, channels of spillovers, are exceedingly robust, some features of the legal system make a patent a less than ideal vehicle for communicating technical information in a timely way despite the requirement that it be written to enable a person of ordinary skill in the art to practice the invention. First, a patent is written by an attorney or a patent agent to persuade an examiner to grant and a court to uphold a property right of the desired scope. Beyond the minimum disclosure required by the patent statute, the applicant has no incentive to disclose information that would be useful to a potential competitor. Second, there is a delay of indeterminate length, sometimes quite long, between the characterization of the invention and its disclosure in an issued patent or a published patent application.[51] Undoubtedly, however, the ability to file for patent protection permits the early communication of inventions through the other sources noted above long in advance of the corresponding patent's publication.

There are nevertheless some features peculiar to the U.S. patent system that concern the committee. The first has to do with the publication of patent applications. As part of the international Trade-Related Aspects of Intellectual Property Rights (TRIPS) agreement concluded in 1994 the United States acceded to the

[50]Saxenian (1996) contrasts the relatively free communication across corporations in Silicon Valley with the relatively restricted communication in Boston's Route 128 high-technology region, a function of differences in laws governing employer-employee nondisclosure agreements.

[51]The delay in publication is determined by statute, 35 U.S.C. 122(b).

general practice elsewhere of publishing patent applications after 18 months.[52] This was recognition that large numbers of U.S. applications became public anyway as a result of foreign filing. But along with the 20-year patent term from first application, it also was intended to foil the practice of "submarine" patenting, whereby an applicant could continue prosecuting a patent in secret indefinitely until it was worth having a patent issued to sue an unsuspecting infringer. Publication has the added benefit of making the technical information available earlier, sometimes considerably earlier, than would otherwise be the case. For applications that never result in patents, publication makes available information that might not otherwise be disclosed at all.

Congress, in implementing the agreement, responded to complaints of some independent inventors that early disclosure of their inventions would expose them to predatory behavior by large companies. The legislation left applicants an option to maintain the secrecy of their applications if they declared that they did not intend to seek protection in any country other than the United States. It may be that many of the applications withheld pertain to marginal inventions not seen to be worth patenting abroad, but by sheer numbers of applications, the exclusion is not insignificant. Overall, the withholding rate was just over 11 percent in fiscal year 2002, up slightly from the previous year (see Table 3-1). However, in computer architecture and software, not a patenting domain dominated by small entities, the opt-out rate was 18.2 percent. Biotechnology and chemicals and materials had the lowest, but not negligible, rates of withholding applications from publication. This discrepancy is not surprising. More patent applications are being kept confidential in fields with the shortest product cycles.[53]

A second unusual feature of the U.S. legal system that may undermine the utility of patents as sources of technical information is the doctrine of willful infringement. Awareness of a patent subjects an accused infringer to the possibility of having to pay triple the amount of damages awarded by a jury finding infringement. Although the committee has no basis for assessing how prevalent the concern is, in the course of our deliberations a number of corporate presenters, particularly in the information technology sector, claimed that this liability is a substantial disincentive to consulting the patent literature.

---

[52]Some may wonder, especially in light of accelerating technology cycles, why patent applications are not published immediately in the interest of timely disclosure of the technical information they contain. A partial answer is the deeply ingrained notion that patent prosecution is an *ex parte* proceeding (that is, between the applicant and the agency or examiner), appropriately reinforced, at least for a period of time, by secrecy.

[53]Some unknown number of applications, however, are not being filed abroad and thus not published because the invention was made public before an application was filed, but within the U.S. grace period. These inventions, therefore, are made available to others in a timely manner.

TABLE 3-1 U.S. Patent Applications Withheld from Publication, FY 2001 and 2002

	Total Perfected[a] Applications	Applications Requesting Nonpublication	Nonpublication Percentage
FY 2001			
Totals	145,578	14,432	9.9
FY 2002			
Biotechnology and organic chemistry	28,718	1,722	6.0
Chemical and materials engineering	36,482	2,470	6.8
Computer architecture and software	27,786	5,064	18.2
Communications	35,513	4,521	12.7
Semiconductor, electrical, optical systems and components	61,367	5,880	9.6
Transportation, construction, commerce, agriculture	36,041	5,177	14.4
Mechanical engineering, manufacturing, products, and designs	42,197	4,949	11.7
Other	11	0	
Totals	268,115	29,783	11.1

[a]Perfected utility and plant applications filed on November 29, 2000, through September 30, 2002.
SOURCE: USPTO.

Ensuring the Timeliness and Containing the Costs of Decisions

Innovation frequently entails high risk and expense. Patents may help induce the investment by providing the patentee with a means of minimizing one source of risk, free use by others of innovation. But if decisions about whether a patent will be allowed or upheld in a dispute, are long delayed, or if the costs associated with those decisions are very high, that alone may tip the balance against investing in an innovation.

Patent pendency, or the elapsed time between the filing of an application and its abandonment or the issuance of a patent, is often cited as the sole measure of USPTO management efficiency. That is misleading. As described and illustrated by the figure in the accompanying "Patent Primer" (see Appendix A), applicants have substantial although not complete control over how long it takes to process a patent application, and they sometimes endeavor to draw the process out even though, for patent applications filed after 1995, delay reduces the lifetime of the eventual patent. A better measure of USPTO performance is the interval between the filing of an application to the office's first response, known as the "First Office Action (FOA)," commonly accepting some claims but denying others. Like average pendency (in 2001, 24.7 months), the time to FOA (14.4 months in

2001) has been slowly increasing, as one would expect where a slowly expanding, or in some technologies, shrinking workforce is coping with a mushrooming workload.

As is often the case, however, averages conceal trends of greater concern. The averages are being held down by processing times for patents in relatively mature technologies, while the most rapidly advancing fields, where the current state of the art is likely to be surpassed in a matter of months, are experiencing lengthening pendency. Applications covering DNA and RNA segments were on average taking well over three years to process by the end of 2002, up by more than six months in three years. Average pendency for Internet business method applicants increased more than eight months in a similar period of time (see Figure 3-5). And the pattern is not confined to technologies where the USPTO, under criticism, took announced steps to be more conservative in its screening of applications.

In the committee's view it takes an inordinately long time to resolve questions of patent validity, whether administratively or in the courts. For patent re-examinations initiated by third-party challengers in the USPTO,[54] the median length of time between the date of application and the final outcome is 7.54 years (Graham et al., 2003).[55] (See Table 3-2.)

There is a longer time lag for settling patent validity challenges through the courts. For a population of cases litigated between 1989 and 1996, Allison and Lemley (1998) found that the average period between the filing of a patent application and a final ruling on the patent's validity was 12.26 years; the average time between the issuance of the patent and resolution of its validity was 8.61 years.[56] (See Table 3-3.)

In these respects the U.S. patent system performs no worse and in some cases better than its European and Japanese counterparts. Average pendency periods were an astonishing 21.4 months shorter in the United States than in the European Patent Office in 2001 and 3 months shorter than in the Japanese Patent Office. The average times to first office action were 14.4 months in the USPTO, 20.7

[54]The types of re-examinations are described in Chapter 3 and in Appendix A.

[55]This estimate is for cases involving patents filed before 1991 to minimize the effects of lag truncation.

[56]This length of time may be influenced by the fact that under 35 U.S.C. 286, damages for patent infringement may be accrued for six years after patent issuance before a suit is filed. Allison and Lemley observe that most patents litigated to judgment involve fairly old technologies on which the patents have existed for some time before they are challenged or enforced. They infer that many firms patent with no immediate intention of enforcing their rights but rather to fence out potential competitors. An alternative explanation is that the rapid developmental pace of some technologies militates against investing the time and resources in lengthy and expensive patent litigation.

TABLE 3-2 Lags in Years Between Patent Application, Grant, Challenge, and Final Outcomes of USPTO Patent Re-examinations Initiated by Third Parties

	USPTO (non-owner requested)		
	No. Obs.	Median	IQ Range
Lag between application and grant	1885	1.75	0.90
Lag between grant and first challenge	1885	2.73	4.81
Lag between first challenge and final outcome	1885	1.42	1.15
Total lag	1885	6.61	5.71
	Pre-1991 Applications Only		
Lag between application and grant	1506	1.80	0.90
Lag between grant and first challenge	1506	3.45	5.68
Lag between first challenge and final outcome	1506	1.42	1.22
Total lag	1506	7.54	6.54

NOTE: The interquartile (IQ) range is a measure of spread or dispersion. It is the difference between the 75th percentile (often called Q3) and the 25th percentile (Q1). The formula for interquartile range is therefore: Q3-Q1. It is sometimes called the H-spread.
SOURCE: Graham et al. (2003).

TABLE 3-3 Lags in Years Between Patent Application, Grant, and Resolution of Validity Challenges in U.S. Litigation

Lag Between Patent Application Filing and Resolution			
	All	Valid	Invalid
Mean	12.26	12.14	12.36
Median	11.3	11.05	11.5
Standard Deviation	5.68	5.29	6.12
Lag Between Patent Issuance and Resolution			
	All	Valid	Invalid
Mean	8.61	8.69	8.49
Median	7.8	8.0	7.5
Standard Deviation	5.08	5.02	5.16
Pendency of Application (time in prosecution)			
	All	Valid	Invalid
Mean	3.64	3.45	3.87
Median	2.7	2.65	2.75
Standard Deviation	2.98	2.56	3.39

SOURCE: Allison and Lemley (1998).

months in the EPO, and 22 months in the JPO (Japan Patent Office et al., 2001). U.S. patent re-examinations take less time to resolve than do challenges in the European patent opposition procedure, given that the window to request an opposition is open for only nine months after a patent issues while a U.S. re-examination may be requested at any time in the life of the patent (Graham et al., 2003).

Application filing fees and fees to maintain patents in force are also lower in the United States than in Japan. It is the cost of legal counsel that puts transaction costs in the United States far beyond the range of those in other industrial countries, and they are rising at a rate much in excess of inflation. The American Intellectual Property Law Association (AIPLA), from its biannual survey of practitioners, estimates that processing a relatively simple U.S.-origin patent application that progresses through examination without amendment or negotiation costs the applicant at least $7,500 in administrative and legal fees in 2002.[57] A complex biotechnology or computer patent subject to multiple amendments could cost tens of thousands of dollars. Albeit with only three data points, 1998, 2000, and 2002,[58] the association estimates that the cost is increasing at an annual rate of 6 to 12 percent (AIPLA, 2003). Estimated costs of various steps in prosecution of different types of patents, compared over five years, are shown in Table 3-4.

The costs of patent conflicts, which almost invariably combine issues of infringement and patent validity, have also increased rapidly, especially for complex lawsuits involving very high stakes, according the AIPLA survey (see Table 3-5).

The median costs to each party of proceeding through a patent infringement suit to a trial verdict are at least $500,000 when the stakes are relatively modest. When more than $25 million is at risk in a patent suit, the median litigation costs for the plaintiff and the defendant average $4 million each, and in the highest stakes, patent suit costs can exceed this amount by more than fivefold. Since relatively few infringement disputes reach trial, almost certainly the more significant transaction costs are the time and attention business managers and counsel spend considering raising a patent challenge, evaluating and responding to others' challenges, devising and carrying out negotiation strategies, and arriving at and implementing settlements.

What is clear is that the burden of costs and uncertainties entailed in challenging and defending patents falls disproportionately on smaller, less experienced

[57]The U.S. General Accounting Office has estimated this minimum cost for an individual or small business paying "small entity" filing and issuance fees at about $6,412, including attorney charges. To maintain the patent for its full term would cost, in addition, approximately $3,528 in fees and attorney costs (U.S. GAO, 2002).

[58]Costs are in nominal dollars, unadjusted for inflation. There are other reasons for caution in interpreting the AIPLA results. The survey has a low response rate and as a consequence may be subject to bias and sampling error.

TABLE 3-4 Increase in the Cost of Prosecuting Patent Applications

U.S. Utility Patents	1998	2000	2002	Percent Change, 1998-2002
Novelty search	$ 999	$1,250	$ 1,500	50.2
Original nonprovisional application on invention of minimal complexity	$4,008	$5,002	$ 5,504	37.3
Provisional application	$2,000	$2,501	$ 2,993	49.7
Original application, relatively complex biotechnology or chemical	$8,000	$9,967	$10,001	25.0
Original application, relatively complex electrical or computer	$7,993	$9,970	$ 9,995	25.0
Original application, relatively complex mechanical	$6,007	$7,996	$ 8,001	33.2
Application amendment or argument of minimal complexity	$1,000	$1,200	$ 1,499	49.9
Application amendment or argument, relatively complex biotechnology or chemical	$1,999	$2,499	$ 2,806	40.4
Application amendment or argument, relatively complex electrical or computer	$1,995	$2,497	$ 2,501	25.4
Application amendment or argument, relatively complex mechanical	$1,503	$1,999	$ 2,199	46.3
Issuing an allowed application	$ 302	$ 400	$ 499	65.2

SOURCE: AIPLA (2003).

firms. For example, Lanjouw and Schankerman (2003), in a paper prepared for this project, find large economies of scale in resolving patent disputes. Having a large patent portfolio significantly reduces the probability of filing a suit on any individual patent, conditional upon its observed characteristics. For a small domestic company with a portfolio of 100 patents, the average probability of litigating a given patent is 2 percent. For a larger company with 500 patents, the probability drops to 0.5 percent, a quarter of the rate for smaller firms. The disadvantage borne by individuals and small firms extends to settlement of patent suits out of court. Large firms with substantial portfolios more readily and more quickly settle their infringement disputes. Cohen and colleagues (2000) also find that research and development managers in large firms report patents to be more

TABLE 3-5 Estimated Median Litigation Costs for Each Party of Litigation (thousands of dollars)

	2001	2003	Percent Change, 2001 to 2003
Less than $1 million at risk			
End of discovery	$250	$290	16.0
Inclusive of discovery, motions, pretrial, trial, post-trial, and appeal	$499	$500	0.2
$1-$25 million at risk			
End of discovery	$797	$1,001	25.6
Inclusive of discovery, motions, pretrial, trial, post-trial, and appeal	$1,499	$2,000	33.4
More than $25 million at risk			
End of discovery	$1,508	$2,500	65.8
Inclusive of discovery, motions, pretrial, trial, post-trial, and appeal	$2,992	$3,995	33.5

SOURCE: AIPLA (2003).

effective in protecting the competitive advantage derived from their innovations than do small firms' respondents; and outside the pharmaceutical industry, small firms disproportionately report that the expected cost of defending patents dissuade them from patenting altogether.

Accessing Technologies for Research and Development

In a variety of contexts the feasibility and terms of access to patented technology, usually by means of licenses, are crucial to further research, technology development, commercialization, and diffusion of new technologies, for example,

- cross-licenses on the myriad elements in semiconductor devices, without which multi-billion dollar investments in fabrication operations would not occur or could be held hostage;
- pooled licenses to technologies underlying technical standards permitting interoperability of electronic and communications equipment;
- licenses to multifunctional research tools that are crucial to progress in biomedical research.

Concerns about access to patented technology, whether from the perspective of innovation or competition, tend to be quite specific to industries and firms. We would have a better general understanding of how markets for technology arise, how they work, under what circumstances impediments to innovation arise, and how they could be reduced if we had data on patent-related licenses, but so far, disclosure and data collection are very limited.[59] Evidence has for the most part been limited to anecdotes, case studies, and occasional court cases.

In all of the panel's deliberations there was only one area—biotechnology research and development, primarily where applied to human health—where it was repeatedly suggested that there might be a significant problem of access to patented technology. This is obviously a field of great public interest. It is also a priority of the scientific community, medical products industries, and clinicians to sustain the remarkable productivity of biomedical research and to achieve its promise to yield highly beneficial and lucrative therapeutic and diagnostic products. The role of intellectual property in promoting and perhaps in some instances impeding this progress has already been the subject of a National Academies' public workshop (NRC, 1997) and an aspect of several studies (Institute of Medicine, 2003; Institute of Medicine, forthcoming), and it has received attention from many other organizations (Nuffield Council on Bioethics, 2002; United Kingdom Royal Society, 2003; and Korn and Heinig, 2002).

As we described in Chapter 2, three concerns have been articulated. The first concern, stated in general terms by Merges and Nelson (1990) and Scotchmer (1991) over a decade ago, is that patents on upstream discoveries, if sufficiently broad in scope, can impede follow-on research and development if access to the foundational intellectual property is restricted. The second concern is specific to biotechnology. In a 1998 *Science* article, attorney Michael Heller and legal scholar Rebecca Eisenberg hypothesized the emergence of what they termed an "anticommons" in biotechnology, which could result if assembling the rights to use the numerous separate patented building blocks necessary to pursue a particular line of research or product development proved to be prohibitively costly and time consuming or simply impossible, causing a promising prospect to be avoided or abandoned. The authors speculated that the diversity of players with different objectives and commercial experience—university administrations, research faculty, biotechnology research firms, large pharmaceutical companies, and government laboratories—increased the likelihood that gridlock would occur. Some might overvalue their upstream research tool inventions from the perspective of downstream product developers faced with the enormous costs of bringing medical products to market. Others might insist on conditions (for example, reach-through rights, downstream royalties) unacceptable to potential licensees. The third concern is specific to university and other nonprofit sector research per-

[59]Publicly held corporations must report to the Security and Exchange Commission licensing relationships "material" to their financial performance. Some universities have disclosed licensing data to researchers.

formers. It is that they could be more adversely affected by the potentially high cost of competing in this arena.[60]

Faced with these conjectures and a few anecdotes, the committee decided to take the unusual step of initiating a modest interview-based survey of firms, intellectual property practitioners, researchers, and government personnel to derive the first empirical data on whether any of these conditions is occurring or emerging. Drawing upon approximately 70 interviews with people in all of these categories, Walsh and colleagues (2003) found that the preconditions for these results appear to exist. More than in the past, therapeutic products tend to be associated with multiple patents; and public research institutions, the locus of many upstream discoveries, are patenting and licensing more aggressively. With important caveats, however, the authors do not find that these developments are yet impeding research and drug development in a significant way. This is in part because the number of patents required for most R&D projects remains manageable and in part because the various players have improvised arrangements or followed norms that mitigate the intellectual property complexities that exist.

What the authors term "working solutions" include, as one would expect, negotiated licenses and royalty payments.[61] Patents are also circumvented by inventing around them, using substitute research tools, and locating research activity offshore. Institutional responses include the National Institutes of Health guidelines encouraging research grantees to facilitate access to patented research tools and the steps taken by several research organizations to place results in the public domain, where they become patent-defeating prior art.

According to many university and corporate respondents to the survey, one of the most pervasive working solutions is infringement of patents, especially on tools of precommercial laboratory research, in some cases on the presumption that research is legally shielded from infringement liability by a "research exception," and in other cases on the assumption that patent holders will not sue over research uses. In particular, there is a widely held belief that private firms will not sue university investigators over patent infringement because there is little to be gained financially and a high risk of adverse publicity.

The first caveat concerns access to patented research tools that are keys to progress in one or more broad therapeutic areas and "rival-in-use," that is, they are tools that are primarily used to develop innovations that will compete with one another in the marketplace. Holders of intellectual property on nonrival

[60]Iain Cockburn (2004) speculates that "more and stronger" patents could not only hinder research but ultimately make the pharmaceutical industry less productive and its products more costly by inducing excess upstream entry and making contracting more difficult between biotechnology tool companies and pharmaceutical producers.

[61]The Cohen-Boyer recombinant DNA technique is an example of a nonrival-in-use research tool patent widely licensed at a reasonable cost.

research tools tend to charge prices that permit broad access, and frequently charge lower prices to university researchers who intend to use the tools for non-commercial purposes. But when tools are rival-in-use, it is in the interest of owners either to exploit the technology themselves or grant exclusive licenses. The concern here is that when such tools are important inputs into the discovery and development of commercial therapies, and there exists uncertainty about the best way to pursue a given application—no less a range of applications—no one firm's efforts at downstream development are likely to realize the full potential of the tool. This is because no one firm is likely to see or be able to develop all the different ways that the discovery might be exploited.

One example of exclusive access to a foundational discovery that has raised concern—where it has been argued that broad rather than exclusive access to the discovery would better serve society's interests—is Geron's exclusive rights to pursue human embryonic stem cell research for three cell types. Other cases of rival-in-use patented technologies that are potentially important inputs into the discovery and development of therapies where exclusive use or licensing has raised similar concerns are described in the sidebar (see Box 3-1). We are not suggesting that these cases represent inappropriate exploitation of the technologies involved. The cost and risk of the technologies' development would need to be considered. But they do illustrate the kinds of access issues that arise.

Although there may be only a few identified controversial instances where restricted access may potentially impede subsequent discovery and development, the consequences for research and medicine of even a rare such occurrence could be large. On the other hand, neither is it clear that less exclusive, low-cost access would on balance serve society's interests if such access dampened the incentive to develop the research tools from the outset. At this point we can say that concern about access to them is not misplaced.

A second caveat relates to university researchers' use in clinical research of diagnostic tests involving patented technologies. Merz, Cho, and their colleagues (2002) have conducted several studies of the impact on clinical laboratories of royalty rates on patented tests, infringement claims, and refusal to license some tests at all. One study found that 25 percent of laboratory physicians reported abandoning a clinical test because of patents. They also reported royalty rates ranging from 9 percent for polymerase chain reaction to 75 percent for the human chorionic gonadotropin patent. A number of laboratories ceased using the genetic test for hemochromatosis once the patent issued and it was exclusively licensed to SmithKline Beecham. Here, too, the issue is not straightforward because clinical laboratories charge patients or their insurers for conducting diagnostic tests, earning revenue that distinguishes the provision of clinical services from non-commercial research. Further, there has been no evidence that patients lacked access to these tests.

The third important caveat is that one of the most prevalent "working solutions"—knowing or unknowing infringement often done or condoned in the belief

BOX 3-1
Issues of Access to Patented Research Tools in Biotechnology

NF-kB (NF-kappa B)

Laboratories at Harvard College, the Massachusetts Institute of Technology, and the Whitehead Institute for Biomedical Research discovered a cell-signaling pathway called NF-kB in the 1980s and were awarded a patent (U.S. Patent No. 6,410,516) in 2002 that may cover almost every clinical application of this fundamental signaling pathway. The patent was exclusively licensed to Ariad Pharmaceuticals. Although investigators "at academic and not-for-profit institutions conduct[ing] non-commercial research" may continue working with the technology without a license, according to Ariad, commercial entities must obtain a license. Ariad has sold international nonexclusive sub-licenses to Bristol-Meyers Squibb and DiscoveRx Corporation. In addition to one-time and annual license fees, these licenses also include milestone and royalty payments on products based upon the NF-kB pathway. Furthermore, corporations using products sold by licensed companies may also need to obtain additional licenses from Ariad itself.

In 2002 Ariad and the three research institutions sued Eli Lilly, arguing that Lilly's Evista and Xigris products for osteoporosis and sepsis, respectively, infringe upon their patents since the drugs work via the NF-kB pathway. In support of its lawsuit, Ariad cited several peer-reviewed papers written by Lilly scientists. On May 13, 2003, the U.S. District Court for the District of Massachusetts denied Lilly's motions to dismiss and for summary judgment. Ariad has approached some 50 other companies for royalty payments on current or future products that function via the NF-kB pathway (Rai and Eisenberg, 2003).

COX-2 Enzyme

The University of Rochester patented the COX-2 enzyme (U.S. Patent No. 6,048,850), claiming all drugs that inhibit the enzyme and routes for administering such drugs. The university sued Searle/Pharmacia for patent infringement. The U.S. District Court for Western New York dismissed the university's complaint on the grounds that the discovery in the patent was invalid for lack of "written description" and therefore could not support an infringement claim (*University of Rochester v. G.D. Searle & Co., Inc.,* W.D.N.Y., March 5, 2003). The Court of Appeals for the Federal Circuit affirmed that the patent was invalid (*University of Rochester v. G.D. Searle and Co., Inc.*, 358 F.3d 916, *available at* 2004 U.S. App. LEXIS 2458, *69 U.S P.Q.2d 1886 (BNA)* (Fed. Cir. February 13, 2004)).

CD34

Johns Hopkins University was awarded a patent claiming all antibodies recognizing CD34, an antigen found on stem cells but not on more differentiated cells. The patentee awarded an exclusive license to Baxter. A rival firm, CellPro, combined two discoveries, one a method for using selectively binding antibodies to enrich bone marrow stem cells and the other an antibody that binds to CD34 (although in a different class of antibodies and recognizing a different binding site on CD34) to produce a cell separator instrument for use in cancer therapies. CellPro declined Baxter's offer of a $750,000/16 percent royalty nonexclusive license, while other firms accepted these licensing terms. CellPro instead chose to sue to invalidate the patent. CellPro ultimately lost the case, was ordered to pay damages (including willfulness damages, because it was found to have lacked a good faith belief the patent was invalid) and legal fees, and went bankrupt (Bar-Shalom and Cook-Deegan, 2002).

OncoMouse

Harvard University patented a mouse containing a recombinant activated oncogene sequence that permitted it to be employed as a model system for studying cancer and permitting early-stage testing of potential anticancer drugs. The invention was licensed exclusively to DuPont. After years of negotiations, the National Institutes of Health and DuPont signed a memorandum of understanding (MOU) permitting, among other things, relatively unencumbered distribution of the technology among academic institutions, although under specific conditions. Recently DuPont imposed new conditions on academic licensees (for example, barring use of the technology in industry-sponsored research without taking a commercial license) and began asserting its patent against research institutions that have not accepted the new conditions (A. Neighbour. Presentation to the National Cancer Policy Board, Institute of Medicine, April 23, 2002).

Embryonic Stem Cells

The University of Wisconsin received a broad patent on its embryonic stem cell discovery in 1998. Its affiliate, the Wisconsin Alumni Research Foundation (WARF), licensed the technology exclusively to Geron, Inc., to develop the cells into six tissue types that might be used to treat diseases and gave Geron options to acquire rights to other issue types. When Geron sought to extend its rights to 12 other tissue types, WARF sued the company in order to offer licenses to other firms. Geron and

continued

BOX 3-1 (continued)

WARF reached a settlement in January 2002, narrowing Geron's exclusive rights to three cell types, removing its option to acquire other exclusive rights, and granting rights free of charge to academic and government scientists for noncommercial research (Stolberg, 2001; Pollack, 2002).

BRCA1

Myriad patented a test for the gene, BRCA1, linked to breast cancer. It allows licensees to perform the tests provided that no fees are charged and the tests are not used for clinical purposes. It also provides reduced-fee tests ($1,200 versus $2,680) for use in NIH-funded projects. Nevertheless, the firm takes the position that giving test results to patients crosses the line from a research test to a clinical test even if other conditions of the license are observed (Blanton, 2002).

that the research in question was shielded from liability—appears to have been undercut by a decision of the Federal Circuit Court of Appeals,[62] handed down in October 2002 after our survey was completed. Ruling on a claim of a common law research exemption from patent infringement liability, the court in a case brought against Duke University agreed that research "solely for amusement, to satisfy idle curiosity, or for strictly philosophical inquiry" is protected; but it held that the protection does not extend to organized scientific research activity pursued as part of the legitimate business of an institution, whether nonprofit or for-profit. The "business" of a university, according to the opinion, is research, education, and reputation enhancement. A few months later the Supreme Court declined to hear Duke University's appeal, allowing the decision to stand. The case involved circumstances very different from those arousing concern in the research community. The plaintiff is a former Duke faculty member, the field is laser research, and the patented technology is laboratory equipment. Nevertheless, the holding is in no way confined to those facts.

It is difficult to anticipate the effects of this decision. An informal poll of research institutions reported to a September 30, 2002, meeting organized by the Association of American Universities, American Association of Medical Col-

[62] *Madey v. Duke Univ.*, 307 F.3d 1351, *available at* 2002 U.S. App. LEXIS 20823, 64 U.S.P.Q.2d (BNA) 1737 (Fed. Cir. 2002).

leges, Council on Government Relations, and National Association of State Universities and Land Grant Colleges revealed that a number of institutions were receiving more notification letters with respect to patent infringement in the aftermath of the decision.[63] University administrators and legal counsel are uncertain what precautions to take to avoid infringement. An increase in full-fledged litigation against research institutions may be unlikely, but it is clear that investigators and their institutions must now pay closer attention to the intellectual property issues involved in their work, with an attendant increase in its cost.

Reducing Redundancies and Inconsistencies Among National Patent Systems

Although significant international progress has been made on standardizing the length of patent terms, establishing common rules for the publication of patent applications, and reconciling other national differences, important differences in standards and procedures remain among the U.S., European, and Japanese patent systems, ensuring a burdensome redundancy that imposes high costs on users and hampers market integration. With respect to any economically important invention, at least three sets of examiners analyze essentially the same application and search more or less the same prior art.[64] This drives up the costs of obtaining and maintaining worldwide patent protection to a level that can be afforded only by the largest multinational corporations. It is estimated to cost as much as $750,000 to $1 million to obtain comprehensive worldwide patent protection for an important invention, and that figure is increasing at a rate of 10 percent per year (Mossinghoff and Kuo, 1998). Equally important, duplication of effort also impacts all three governments, which are coping with the surge in patent applications with at best slowly growing and at worst reduced resources.

Impeding full reciprocity or mutual recognition, let alone uniform enforcement of patent rights, are a host of subtle and overt differences in approach, procedures, and standards, some of them technology-specific, many of them subject to ongoing negotiations in the World Intellectual Property Organization

[63]The organizations have arranged with the American Association for the Advancement of Science to continue to monitor universities' experience in this regard.

[64]The Patent Cooperation Treaty (PCT), implemented in 1978 under the World Intellectual Property Organization (see description in Appendix A), has created a division of labor chiefly between the industrial countries and nonindustrial countries with limited or no patent examination capabilities by providing for USPTO or EPO advisory searches and, at an applicant's option, examinations that are frequently accepted by developing countries. Such searches and examinations are available among the trilateral patent offices but are often repeated or duplicated, since applicants frequently file simultaneously under the PCT and in national offices in major markets. The PCT offers applicants an efficient means of filing applications in multiple countries.

(WIPO).[65] Among the principal substantive differences in patent law are the following:

- *Priority of invention.* As between two true inventors—as contrasted with copiers—every nation in the world except the United States provides patents to the inventors who first undertake to use the patent system to disclose their inventions and gain protection.[66] This is conventionally known as the first-inventor-to-file system of priority. The United States provides a patent to the first person to "conceive" and/or "reduce the invention to practice" (first-to-invent system). The latter gives rise to a number of priority disputes, known as "interferences," over the timing and identity of invention that are difficult to adjudicate, whether administratively in the patent office or in the courts. The U.S. system nevertheless has strong adherents among individual inventors and small companies.
- *Best mode requirement.* U.S. law requires that a patent application disclose the "best mode" of implementing an invention to prevent the applicant from concealing the invention's significance by describing a trivial or remotely related application. No other country has such a requirement. The best mode requirement is frequently raised as a defense in patent infringement litigation. In other words, an accused infringer asserts that the patent should be invalidated because of the patent owner's failure to disclose the best mode. Judicial inquiries into best mode require access to inventor records and testimony that are often inconclusive.
- *Grace period.* Under U.S. law inventors can disclose their inventions publicly or commercialize them before filing patent applications as long as the applications are filed within one year. The grace period encourages early disclosure, for example, of research results in scientific publications or conferences, or commercialization of an invention without causing inventors to forfeit their rights to protection. Japan has a more limited grace period in time and scope; Europe provides none.

Maintaining a Level Field Among Rights Holders

Uniform application of the patent law's rights and obligations was not questioned until the U.S. Supreme Court, in June 1999, struck down a federal law that had denied a state from claiming immunity under the Eleventh Amendment of the Constitution when sued in federal court for patent or other federal intellectual property infringement. In *Florida Prepaid Postsecondary Education Expense Board v. College Savings Bank*[67] the Court said that Congress had not shown

[65]WIPO in 2000 resuscitated a set of substantive patent law harmonization negotiations, commonly known as "deep harmonization," that had been quiescent since 1993.

[66]In January 1998 the Philippines abandoned the first-to-invent system, leaving the United States alone in adhering to it.

[67]*Florida Prepaid Postsecondary Education Expense Board v. College Savings Bank.* 527 U.S. 627 (1999).

such a pattern of state agency infringement or an absence of state remedies that would have justified a removal of immunity. As a result of the decision a public university could be in the position of asserting its patent rights against an alleged infringer while successfully barring a patent holder from recovering damages for its own past infringement. *College Savings Bank* does not prevent the patentee from enjoining future use of the patented invention. Still, the partial immunity is not available to a private party, including other universities in the same state. As a result, it could have distorting effects. For example, if investigators in a state institution used a patented research tool one time without license to find a profitable pharmaceutical product, the patentee could sue for an injunction to bar future use of the tool, but it would be pointless. Sovereign immunity prevents the patent holder from suing for past damages, even if they turned out to be substantial.

Like many other issues arising from recent policy changes, it is not clear how serious a problem the disparity represents. It is not enormous. A state could not set up a systematic program of infringement, for example, to produce low-cost prescription drugs for its Medicaid patients. It could be enjoined in federal court and also sued in a state court for an unconstitutional taking of property. Furthermore, if states began to infringe patents systematically, Congress would have the factual predicate the Supreme Court said was necessary to support a waiver of immunity in federal court. Nevertheless, the committee believes the disparity created by the decisions is not negligible. It puts the United States in the position of being out of compliance with the TRIPS agreement, which provides no exceptions for subunits of government. Further, it may over time affect the choices private firms make in supporting research at public or private institutions.

In an analysis for the Senate Judiciary Committee the U.S. General Accounting Office (GAO) reported in 2001 that before the Supreme Court's decision, state entities were rarely sued in federal court for patent or copyright violations; there had been 58 cases during a 15-year period, less than 0.05 percent of the total number of cases in federal district courts. On the other hand, two-thirds of state universities responded to the GAO that they had received accusations of infringement, usually in the form of cease-and-desist letters, during the same period. Seven of nine institutions responding reported receiving 11-15 complaints, and one institution reported receiving more than 16 complaints. Almost certainly, the number of complaints of university infringement and conceivably the number of lawsuits will increase in the aftermath of the *Madey v. Duke* ruling that universities in general may not claim a research exemption defense under common law. On the one hand, private university administrations may conclude that they need to make a much more vigorous effort, which could be burdensome for researchers, to guard against infringement suits than do public university administrators. On the other hand, there may develop a perception that private institutions are more reliable partners in collaborative activities with industrial companies.

SUMMARY

The committee concludes that the U.S. patent system, while functioning reasonably well in many respects, most importantly in its rapid accommodation to technological changes and its flexibility in dealing with differences between technologies, is exhibiting a number of characteristics requiring attention and improvement.

- Although it is not clear that the quality of most patents has declined significantly, there are reasons to be concerned about whether many patents in leading-edge technologies that are drawing substantial investments represent desirable degrees of novelty, utility, and non-obviousness. This appears to be a function both of pressures on the examination process and of interpretations of some patent standards.
- There are remediable features of the U.S. patent system that undermine its function in disseminating technical information.
- Delays and costs entailed in resolving questions of patentability, the validity of issued patents, and infringement, although in some respects comparing favorably to those in Europe and Japan, excessively compound the uncertainty surrounding innovation.
- Difficulties accessing the patented technology necessary to sustain the progress of biomedical research and therapeutic product development have in some cases raised the cost and modified the character of research and in a very few instances have become a serious obstacle. This may become a more significant problem with the greater complexity of research and proliferation of patents on technologies well upstream of commercial products, and in the aftermath of a recent federal appeals court decision denying fundamental research protection from patent infringement liability.
- Although progress has been made in harmonizing national patent systems, substantial differences in procedures, standards, and substantive law remain and impede achieving reciprocity or mutual recognition of patent search and examination results among the United States, Europe, and Japan.
- In interpreting the Eleventh Amendment to the Constitution, the United States Supreme Court recently raised a troublesome disparity between state and private institutions with respect to their obligations under federal intellectual property law.

4

Seven Recommendations for a 21st-Century Patent System

The committee supports several steps to ensure the vitality and improve the functioning of the patent system.

- ***An open-ended, unitary, flexible patent system.*** The system should remain open to new technologies with features that allow flexibility in protecting new technologies. Among the features that should be exploited is the United States Patent and Trademark Office's (USPTO) development of examination guidelines for new or newly patented technologies. The office should seek advice from a wide variety of sources and maintain a public record of the submissions in developing such guidelines, and the results should be given appropriate deference by the courts. The Court of Appeals for the Federal Circuit ("Federal Circuit") also should ensure its exposure to a variety of expert opinions by encouraging submission of amicus briefs and by exchanges with other courts. In addition to qualified intellectual property professionals, appointments to the Federal Circuit should include people familiar with innovation from a variety of perspectives—management, finance, and economics, as well as nonpatent areas of law affecting innovation.
- ***Non-obviousness standard.*** The requirement that to qualify for a patent an invention cannot be obvious to a person of ordinary skill in the art should be assiduously observed. In an area such as business methods, where the common general knowledge is not fully described in published literature that is likely to be consulted by patent examiners, another method of determining the state of general knowledge needs to be employed. Given that patent applications are examined *ex parte* between the applicant and the examiner it would be difficult to bring in other expert opinions at that stage. Nevertheless, the Open Review procedure

described next provides a means of obtaining expert participation after a patent issues. With respect to gene-sequence-related inventions, a low standard of nonobviousness results from Federal Circuit decisions making it difficult to make a case of obviousness against a genetic invention (for example, gene sequences). In this context the court should return to a stricter standard, which would also be more consistent with other countries' practices in biotechnology patenting.

- ***Open Review procedure.*** Congress should seriously consider legislation creating a procedure for third parties to challenge patents for a limited period after their issuance in an administrative proceeding before administrative patent judges of the USPTO. The speed, cost, and design details of this proceeding should make it an attractive alternative to litigation to determine patent validity and be fair to all parties.
- ***USPTO capabilities.*** To improve its performance the USPTO needs additional resources. These funds should enable hiring additional examiners, implementing a robust electronic processing capability, and creating a strong multidisciplinary analytical capability to assess management practices and proposed changes. In addition, the funds should be used to provide early warning of new technologies being proposed for patenting, and to conduct reliable, consistent, reputable quality reviews that address office-wide as well as subunit and examiner performance. The current USPTO budget does not suffice to accomplish these objectives and to administer an Open Review procedure.
- ***Research liability for patent infringement.*** In light of the Federal Circuit's 2002 ruling that even noncommercial scientific research enjoys no protection from patent infringement liability, and in view of the academic research community's belief in the existence of such an exemption, and behavior accordingly, there should be some level of protection for noncommercial uses of patented inventions. Congress should consider appropriately narrow legislation, but if progress is slow or delayed the Office of Management and Budget and the federal government agencies sponsoring research should consider extending "authorization and consent" to grantees as well as contractors, provided that such rights are strictly limited to research and do not extend to any resulting commercial products or services. Either legislation or administrative action could help ensure preservation of the "commons" required for scientific and technological progress.
- ***Litigation elements.*** Three provisions of patent law that are frequently raised by plaintiffs or defendants (rarely by the courts) in infringement litigation depend on determining a party's state of mind, and therefore generate high discovery costs. These provisions are (1) "willful infringement," which if proven, exposes an infringer to possible triple damages; (2) the doctrine of "best mode," which addresses whether an inventor disclosed in an application what the inventor considered to be the best implementation of the invention; and (3) the doctrine of "inequitable conduct," concerning whether the applicant's attorney intentionally misled the USPTO in prosecuting the original patent. To reduce the cost and

increase the predictability of patent infringement litigation outcomes, and to avoid other unintended consequences, these provisions should be modified or removed.

- ***International harmonization.*** The United States, Europe, and Japan should further harmonize patent examination procedures and standards to reduce redundancy in search and examination and eventually achieve mutual recognition of results. Differences that among others are in need of reconciling include application priority ("first-to-invent" versus "first-inventor-to-file"), the grace period for filing an application after publication, the "best mode" requirement of U.S. law, and the U.S. exception to the rule of publication of patent applications after 18 months. This objective should be pursued on a trilateral or even bilateral basis as well as a multilateral basis.

Although some of our recommendations parallel those of previous commissions and reports, the most relevant comparison is with the proposals of the Federal Trade Commission (FTC) in its report released in October of last year. Although we approached the operation of the patent system from different perspectives, addressed somewhat different topics, and employed quite different methodologies, there are several areas of agreement.

- The USPTO and the Court of Appeals for the Federal Circuit should broaden their consideration of relevant economic and technical analysis.
- The non-obviousness standard should be more vigorously applied, at least in some technological fields.
- Congress should create a review procedure for challenging and reviewing issued patents.
- The financial resources of the USPTO should be increased.
- All patent applications should be published after 18 months.
- The legal doctrine subjecting "willful" infringers to enhanced damages should be modified or eliminated.

PRESERVE A FLEXIBLE, UNITARY, OPEN-ENDED PATENT SYSTEM

Innovation processes differ markedly from one industrial sector to another. There is ample evidence that development lead times, product cycles, the relative dominance of cumulative or interoperative or stand-alone innovations, capital investment requirements, and even sources of innovation all vary greatly. We know, too, that firms in different industries acquire, value, and exercise patents differently. Accordingly, the optimal number, coverage, and division of patent rights to encourage innovation may vary. These circumstances, some might argue, call for designing a formal (that is, statutory) system in which patent standards, strength, duration, and other features vary from technology to technology and, conceivably, certain technologies are excluded from patenting altogether.

Historically, there has been strong resistance to a differentiated patent system and to subject matter exclusions and fairly consistent adherence to a relatively open-ended unitary system. Exceptions, although more common recently, are relatively few and narrow and usually in the nature of limited exceptions rather than sui generis systems of intellectual property protection. For example, in 1996 Congress exempted medical practitioners and related health care entities using patented medical procedures from infringement liability rather than bar surgical procedure patents altogether.[1] It lengthened terms for some pharmaceutical patents to compensate for regulatory delays[2] and protected certain experimental uses of pharmaceuticals by generic suppliers from liability.[3] Recently, Congress was persuaded that the advent of business method patents might snare longtime users of newly patented business procedures in infringement suits; but, rather than curtail the issuance of such patents or limit their terms, legislation made prior use a defense available to accused infringers of that class of patents.[4] A special obviousness provision deals with concurrent process and composition-of-matter claims on biotechnology patent applications.[5] The Plant Patent Act of 1952[6] and the Plant Variety Protection Act of 1970,[7] representing modified patent regimes, and the 1984 semiconductor mask protection legislation[8] and the 1988 Vessel Hull Design Protection Act,[9] representing modified copyright regimes, are the only examples of new statutory classes of intellectual protection designed for particular technologies.

Apart from the very recent congressional ban on human organism patents,[10] clearly a special case, there have been no successful legislative attempts to circumscribe patenting. Some members of this committee are concerned about patent incursions on the public domain of ideas and information, particularly in the realm of scientific research results. Even so, they believe that the proper approach is on a case-by-case basis through a mechanism for review of issued patents for conformity with the statutory standards and associated case law rather than an attempt to draw the line more sharply. The entire committee endorses such a postgrant review procedure, described later in this chapter.

The committee also agrees that given the state of our knowledge there are strong reasons to preserve a formally unitary system. For one thing, we do not

[1]35 U.S.C.§ 287(c) (2000); P.L. 104-208, 1996 HR 3610.
[2]35 U.S.C. §§ 155, 156 (2000).
[3]35 U.S.C. § 271 (2000).
[4]35 U.S.C. § 273(b)(1a)(3) (2000).
[5]35 U.S.C. § 103(b) (2000).
[6]35 U.S.C. §§ 161, 164.
[7]7 U.S.C. § Sec 2321 *et seq.*
[8]17 U.S.C. § 901-14 (2000).
[9]Part of the Digital Millennium Copyright Act, P.L. 105-304.
[10]It has been USPTO policy since 1987 not to issue any human organism patents.

know enough about innovative processes to advise Congress on the optimal characteristics of different classes of patents in different circumstances. Legislative tailoring of the patent system to each major industry would be a prime opportunity for interest-group politics to influence the results, which could be quite resistant to change or adjustment. Even if Congress were able to get it "right" in economic terms, technological change and industries' structural evolution might render the specifications obsolete and possibly counterproductive. That appears to have been the case with semiconductor mask protection. Although the industry lobbied vigorously for the legislation, and there have been a number of filings under it, few in the industry still view it as an important way to protect proprietary chip design, primarily because the underlying technology evolved rapidly, obviating the perceived need. The fact that such instances are rare suggests that Congress has no great appetite for crafting industry- or technology-specific patent policies. In any case, it has largely tied its hands by ratifying the TRIPS agreement of 1994, which prohibits signatory states from discriminating in the grant of patents based on the technology involved.[11]

The committee realizes that there may appear to be some contradiction between this position and our belief in the importance of exploiting the mechanisms and doctrines that reflect differences among technologies or allow for some deliberate discrimination among them by the USPTO, by the courts, and by patent holders themselves. These include subtle differences in the application of the common legal standards of obviousness, enablement, and written description, and the various other policy levers described by Burk and Lemley (2003a). The difference is that these mechanisms, in contrast to legislation, allow for incremental adjustments that are more easily made.

In particular, the committee endorses the USPTO's development of examination guidelines, outlining how it will apply the statutory standards to emerging technologies. In the case of the utility and written description guidelines for genetic inventions and earlier in the case of computer programs, this was accomplished through a notice and public comment process not unlike that employed by federal regulatory agencies in formal rulemaking proceedings. This is not only a means of achieving some degree of standardization in USPTO practice involving a new technology or newly patented technology well before a number of validity cases are decided by the courts. It is also a means of obtaining advice from a variety of sources in a way that is open to all interested parties.[12]

The USPTO should solicit comments from legal scholars, economists, and independent experts as well as stakeholders and maintain an open record of the submissions. Further, the guidelines and the record behind them should be part of

---

[11]Article 27(1).

[12]The USPTO periodically requests comments on issues other than those in proposed rulemaking, and these are publicly available.

the record in any appeal to a court, where they could be used to inform judicial decisions. There are other ways to expose the courts to a wide range of opinion and analysis.

The Court of Appeals for the Federal Circuit is in most instances the final arbiter of patent law. Both students of and practitioners before the court are in general agreement that the 1982 centralization of patent appeals in the Federal Circuit has been a vast improvement over adjudication in the circuit courts of appeals. It reduced forum shopping, focused attention and thought on neglected issues of patent law, produced innovations at the trial court level, and in general yielded greater consistency. At the same time, specialized institutions have insular tendencies.[13] For example, the Federal Circuit appears to rely less on independent scholarly analysis, even legal scholarship, than the regional generalist appeals courts. Nard (2002) found that the circuit courts cite scholarly work roughly four times as often as the Federal Circuit. He acknowledges that the Federal Circuit is more familiar with patent law than regional courts are with, say, copyright or trademark law; but he suggests that the disparity is such that the "court verges on the abstract by failing to give adequate weight to empirical and economic scholarship."[14]

We recommend some modest steps to ensure that the Federal Circuit, despite its specialization, has broader exposure to legal and economic analysis in all areas of innovation-related law and to the impact of its decisions on the lower courts and on the Patent Office.

- ***Briefs.*** The Federal Circuit should encourage the submission of briefs that draw upon insights from other judicial decisions, legal scholarship on the patent system, and the growing body of patent-related economics literature. In particular,

[13]The definitive analysis of specialized adjudicative bodies and their biases is by Marver Bernstein (1955). Interestingly, some of the early patent appeals court proposals did not contemplate permanent appointments to the court. For example, the National Research Council (1919) recommended selection by the chief justice from the district and circuit court benches, with service limited to a six-year term unless reappointed. The 1936 National Research Council report recommended permanent appointment, but of judges with diverse scientific and technical backgrounds as well as experience in the trial of patent cases.

[14]Judge Paul Michel, speaking at the University of California, Boalt Hall, Conference on Patent System Reform, March 1, 2002, made a similar point about the recent litigation on the issue of the doctrine of equivalents in *FestoCorp. v. Shoketsu Kinzoku Kogyo Kabushiki Co.*, 234 F.3d 558, 56 U.S.P.Q.2d 1865 (Fed. Cir. 2000) (en banc), overruled in part by 535 U.S. 722, 122 S. Ct. 1831, 152 L. Ed. 2d 944 (2002): "Now you might have thought . . . where there was a concern about the relative needs to promote adequate incentives—or you could say fairness to inventors—on the one hand with the need for competitors to have adequate predictive value and certainty on the other hand, that somebody at least amici and, one would hope, also the parties would have given us some very meaningful data about that. Now I read all the briefs . . . and I can't remember anything that I would consider empirical data. . . . If you trace back the pedigree I suspect that you will find that in a great many cases there never was any meaningful economic or quantified data."

as the court has done in some recent notable cases,[15] it should welcome amicus briefs, as these tend to raise broader issues and cite a wider range of literature than do the briefs of parties to cases.

- ***Appointments.*** Because the Federal Circuit's docket extends to diverse cases far removed from patent or, more broadly, innovation law, the few appointments intended to support the court's expertise in that area should be made with particular care. They should not be confined to intellectual property practitioners and academics. Rather, the court's perspective should be broadened by appointing judges familiar with innovation more generally, including men and women with backgrounds in antitrust or finance law or, in addition to their legal training, in economics or economic history. Furthermore, some appointments might be through the elevation of regional district court judges, a routine practice with respect to the regional appeals courts but not the Federal Circuit. The addition of one or more district judges with patent litigation experience would give the court perspective not only on problems at the trial court level but also on economic issues outside its jurisdiction.
- ***Designations.*** Trial court judges are often asked to sit by designation in other courts of appeals, helping the system as a whole keep track of jurisprudential trends. Federal Circuit judges have rarely participated in this practice.[16] The committee suggests that the Federal Circuit invite regional judges to sit on its panels and regional circuits invite Federal Circuit judges to sit by designation. This would give Federal Circuit judges a better sense of how patent law fits in with other laws influencing innovation and how other courts incorporate economics into their decision making.

REINVIGORATE THE NON-OBVIOUSNESS STANDARD

Non-Obviousness and Business Method Inventions

The non-obviousness determination is necessarily a judgment, not something that can be resolved through a bright line test. Fundamentally, it assumes that an invention is novel and the decision maker must determine whether the hypothetically skilled person in the art would nonetheless have considered the novel invention something within the routine skill of the field. The USPTO and the reviewing courts are concerned that an invention that was genuinely non-obvious before it was made may often look obvious in retrospect. The courts have been vigilant in

[15]For example, the Federal Circuit recently granted an *en banc* review to consider the willfulness doctrine discussed below and actively solicited amicus briefs on a wide range of issues to aid in its deliberations. *See Knorr-Bremse Systeme Fuer Nutzfahrzeuge Gmbh v. Dana Corp.* 344 F.3d 1336 (Fed. Cir. 2003).

[16]See Dreyfuss (forthcoming) for suggestive data on the extent to which circulation of judges into and out of the Federal Circuit occurs.

preventing hindsight based on the inventor's patent disclosure from leading to an "obvious" determination. The doctrines that protect against hindsight, while well founded in most other contexts, assume that the USPTO will have access to the state of the art at the time the invention at issue was made. That assumption has not been true in the business methods area.

The prior art by which inventions are judged is defined by 35 U.S.C. § 102. The most common forms of prior art in conventional areas of technology are printed publications, including scientific journals and patents. The prior art also includes prior information "known or used by others." In foreign patent systems this is known as the "common general knowledge," which describes the concept somewhat better. Under U.S. and foreign law it is not enough that one or a few people in the field have the information alleged to be prior art. The information must be generally known to be patent defeating. In conventional technologies the published literature in fact represents a fairly good catalog of the common general knowledge in the field, and the USPTO can therefore access it. That is not the case for business methods, which may not be written in any of the places likely to be consulted by examiners.[17]

Several examples illustrate how the conventional approach to obviousness breaks down in the business methods context. For example, one common form of patent application involves use of a new component or ingredient to replace an old one for a particular function. In patent law this is an issue of "equivalence," that is, whether the new component is recognized in the prior art as an equivalent of the old component. Section 2144.06 of the Manual of Patent Examining Procedure (MPEP) addresses this issue, relying on three Federal Circuit cases. It states, "In order to rely on equivalence as a rationale supporting an obviousness rejection, the equivalency must be recognized in the prior art, and cannot be based on applicant's disclosure or the mere fact that the components at issue are functional or mechanical equivalents." In other words, use of a new ingredient, even a familiar ingredient in other circumstances, can lead to a patentable invention unless the USPTO can determine that such a substitution was part of the state of the art. This understanding in the art can be based on the common general knowledge apart from the published literature.[18] In an area like business methods, where the published literature does not fully describe the state of the art, the USPTO is

[17]For example, relevant prior art, such as for the patent on the inverse elasticity rule, appears in economics texts. But much of the prior art for business methods is embodied in services and processes not described in any literature (Laurie and Beyers, 2001).

[18]*Graham v. John Deere Co.,* 383 U.S. 1 at 10 (1966) quoted Jefferson that "[A] change of material should give no right to a patent. As to the making a ploughshare of cast rather than of wrought iron; a comb of iron instead of horn or ivory . . ." T. Jefferson, Letter to Isaac McPherson (Aug. 1813), VI Writings of Thomas Jefferson, at 181. Presumably this statement was not based on a learned treatise suggesting cast-iron ploughs but the common understanding at the time that cast iron was a general substitute for wrought iron in the latter's many applications.

severely handicapped in its ability to make rejections based on the obviousness of a substitution.

A corollary to the common-general-knowledge principle involves the combining of ideas from different sources. Every invention at some level is a combination of old elements. Again, experience has shown that truly non-obvious inventions will be denied patent protection unless the decision maker guards against the use of hindsight.[19] USPTO policy reflects this concern. Based on Federal Circuit and Court of Customs and Patent Appeals (CCPA) holdings, MPEP 2143.01 states that "[t]he mere fact that references *can* be combined or modified does not render the resultant combination obvious unless the prior art also suggests the desirability for the combination." Again, the poverty of the published literature on business methods makes it difficult for the USPTO to make obviousness rejections.

If the problem of the non-obviousness standard's application to business method inventions is lack of access to published information regarding the common general knowledge in the field at a particular time, that knowledge is best provided through testimony (affidavit or live) by those active in the field at the relevant time. Consideration should therefore be given as to how the USPTO could obtain such testimony.

Patent applications are examined *ex parte* in secret between the applicant and the USPTO. The USPTO does not employ experts to provide evidence that might support an obviousness rejection. One approach would be to change this practice to admit the testimony or written opinion of the USPTO-appointed experts. This would, however, prolong the patent prosecution and make it more expensive. Perhaps more importantly, there are significant concerns regarding the USPTO's ability to maintain the impartiality it should have respecting the merits of an application. The examiner already plays the dual role of adversary and "judge" during the examination process. It is likely that the USPTO's testimonial evidence will be contradicted by testimony submitted by the applicant. It may not be wise to have the USPTO act as the decision maker resolving the competing evidence where one of the sources is an expert retained by the USPTO itself. This concern is particularly compelling in view of the recent Supreme Court decision, *Dickerson v. Zurko*,[20] which held that on appeal the courts will have to affirm any

[19] An example is *In re Fine*, 837 F.2d 1071 (Fed. Cir. 1988), which involved a system for detecting and measuring certain nitrogen compounds by using a gas chromatograph, a converter to oxidize the nitrogen compounds into nitric oxide, and a nitric oxide detector. Two previous references were relevant; one disclosed an analogous approach to monitoring certain sulfur compounds, and the other described nitric oxide detectors. Although the examiner and the Board of Patent Appeals thought it obvious to substitute the nitric oxide detector in the system, the Federal Circuit found that there was no support for such a conclusion. The USPTO had not demonstrated that the idea of the combination of teachings was within the state of the art.

[20] 527 U.S. 150 (1999) which reversed and remanded *In re Zurko*, 142 F.3d 1447 (Fed. Cir.1998).

USPTO finding of fact (for example, what was part of the common general knowledge) unless there was no substantial evidence to support the finding. Merely disagreeing with the USPTO on how the evidence should have been weighed will not be a basis for reversal. The testimony of the USPTO's witness would almost always meet the substantial evidence test.[21] Thus, the use by the USPTO of retained experts during examination does not appear to be a workable solution.

Elsewhere in this report the committee has proposed the implementation of an efficient process in the USPTO by which third parties could challenge the validity of an issued patent. This process, which we call Open Review, would be a preferable method of bringing testimony regarding the common general knowledge to the attention of the USPTO. The parties affected by the patent will likely be in the best position to obtain testimony from those working the field at the relevant time. The USPTO will not have a vested interest in either side's experts. Further, since obtaining and evaluating testimony requires more resources than conventional patent prosecution, the Open Review forum will tend to confine those costs to cases in which the patent is of some market importance.

It may be argued that the case of business methods is not only unique but transitional and therefore of little broader significance. We are not confident on either score. Some of the apparently obvious patents listed in Chapter 3 appear to have the same characteristic and therefore may have been approved not carelessly but under the prevailing rule that references should not be combined for the purpose of proving non-obviousness unless the examiner can point to a specific piece of prior art that says the references should be combined. The business methods arena lacks a publication culture but even where such a culture exists, scientists, artisans, and creative people generally speaking strive to publish *non*-obvious information. So if it is obvious to those of skill in the art to combine references, it is unlikely that they will publish such information. It is therefore difficult to imagine that another class of patent applications will not pose the same issue in the future. In the meantime, with business methods patent grants, it is true that there will be a steady accumulation of patent prior art. But even it will be limited if the United States remains the only major country issuing business method patents. Moreover, given the great variety of business method applications, the lack of nonpatent published information may remain a significant handicap in assessing obviousness.

---

[21]USPTO reliance on expert witnesses raises additional concerns. One concern is whether the office would have to disclose to the applicant any contrary expert opinions it has obtained; in much the same way the applicant must disclose adverse information to the USPTO under the duty of disclosure. It is also questionable whether the USPTO would have access to those persons most knowledgeable about the state of any art. Such individuals may be competitors of the patent applicant. Thus, it will be problematic to retain them for the secret examination process. These are the types of individuals the applicant wishes to exclude.

Non-Obviousness and Gene Sequence-Related Inventions

One basis for rejecting an invention for obviousness has been to allege that the invention was "obvious to try." This test has been rejected by the courts because it penalizes those who devise a sensible plan of research by exploring the paths most likely to succeed. Particularly when success of the chosen path was not assured, eliminating the patent incentive in such a circumstance was recognized as contrary to the purpose of the patent system. The courts, therefore, have held that an invention is only obvious and unpatentable when the obvious route to try is coupled with a "reasonable expectation of success."

One of the earliest Federal Circuit decisions in the field of biotechnology, *In re O'Farrell*,[22] found that an invention related to gene expression was unpatentable under the above test even though success was not assured. The *O'Farrell* court dealt with how high the "reasonable" bar should be set and set it quite low. The court found that the inventors' own prior publications with similar systems expressing "nonsense" sequences in *E. coli* provided a reasonable expectation that actual gene sequences from an exogenous source would be expressed into functional proteins. This decision made the obviousness standard easier to use in rejecting applications.[23] The USPTO has applied this standard for biotechnology in general.[24]

If the early technical advance of *O'Farrell* in the relatively unpredictable period during which recombinant gene expression was still being worked out was obvious under the above standard, it is a fair question to ask why are not most of the gene sequences from the human genome project obvious and therefore unpatentable. After all, the technical question presented by the genome project was not whether the human genome could be sequenced, but which group would finish first. The sequencing of the genome, and other collections of mass sequence-related data (for example, expressed sequence tags [ESTs]) would appear to be obvious to try with a reasonable expectation of success.[25] The reasonable expec-

---

[22] 853 F.2d 894, *available at* 7 U.S.P.Q.2d 1673 (BNA) (Fed. Cir. 1988).

[23] The Supreme Court's decision in *Dickerson v. Zurko*, 527 U.S. 150 (1999) should further strengthen the USPTO's position in making obviousness rejections under the reasonable expectation test. Several of the key issues underlying the obviousness test are factual issues, namely the content of the prior art, level of skill in the art, expectation of success, and motivation to combine prior art. Thus, applicants will have a difficult time overturning adverse USPTO decisions if there is any reasonable evidentiary basis to support an expectation test.

[24] For example, the obviousness standard has defeated patentability for novel monoclonal antibodies prepared against known antigens using routine techniques. *Ex parte Erlich*, 22 U.S.P.Q.2d 1463 (BNA) (Bd. Pat. App. & Int. 1992).

[25] The law permits an applicant to overcome a presumption of obviousness if an applicant can demonstrate an unexpected or superior property of the claimed invention not shared by the prior art. *In re Papesch*, 315 F.2d 381, 137 U.S.P.Q. (BNA) 43 (CCPA 1963); *In re Dillon*, 919 F.2d 688, 16 U.S.P.Q.2d (BNA) 1897 (Fed. Cir. 1990). Thus, patentability for a new gene might lie in the discovery of an unexpected or superior property of the gene, or more likely, the protein it encodes. Many

tation test has not been applied because a pair of Federal Circuit decisions dealing with cloning inventions from the 1980s created a de facto rule of per se non-obviousness for a novel genetic sequence.

The *In re Bell* decision is illustrative.[26] The USPTO in that case argued that a defined gene sequence was obvious from prior art, including the sequence of the encoded protein and a general method of cloning. The inventor argued that the prior art relied upon by the USPTO did not suggest all of the modifications to the cited cloning technique to make it operative and that the USPTO had, without supporting evidence, deemed such modifications within the ordinary skill of the field. The Federal Circuit, rather than merely find that the prior art did not provide, for example, sufficient information to make success a reasonable expectation, went one step further. It applied obviousness concepts developed for synthetic chemical compounds.

In *Bell* and then *In re Deuel*[27] the court held that a gene is just another type of chemical compound and the issue for non-obviousness is the structure (that is, sequence) of the gene. Unless the sequence is predictable from the prior art, the gene is non-obvious. The court created a per se rule that the obviousness of obtaining the gene could *never* be relevant to patentability. This per se rule is highly unusual and flies in the face of significant Federal Circuit precedent rejecting the creation of any per se rules relating to non-obviousness.[28]

In the synthetic chemical field the invention usually resides in the design of the new compound. The method of making the compound might be an additional technical hurdle that adds to patentability, but usually a method is obvious once the compound is designed. Thus, the ease or difficulty of making a newly designed compound could add to its patentability but could not defeat patentability of a compound the structure of which was non-obvious.[29] According to current doctrine in synthetic chemistry, the focus of patentability is the non-obviousness of the chemical structure. The fact that if someone had the design they would know how to make the compound has no bearing on patentability.

---

genomics patents, however, only speculate as to usefulness of the novel gene. Such speculation ranges from a virtual "laundry list" of potential applications to a more specific routine comparison of the novel gene to genes of known function. Such speculation is itself obvious and routine and, since it is based on well-known techniques, should not be considered the discovery of an unexpected or superior property.

[26]*In re Bell*, 991 F.2d 781, *available at* 26 U.S.P.Q.2d 1529 (BNA) (Fed. Cir. 1993).

[27]*In re Deuel*, 51 F.3d 1552, *available at* 34 U.S.P.Q.2d 1210 (BNA) (Fed. Cir. 1995).

[28]*In re Ochiai,* 71 F.3d 1565, *available at* 37 U.S.P.Q.2d 1127 (BNA) (Fed Cir 1995); *In re Pleuddemann*, 910 F.2d 823, *available at* 15 U.S.P.Q. 2d 1738 (BNA) (Fed. Cir. 1990); *In re Durden*, 763 F.2d 1406, *available at* 226 U.S.P.Q. 359 (BNA) (Fed. Cir.1985).

[29]There is one situation when the obviousness of how to make a synthetic chemical invention is highly relevant. A patent can be obtained for a compound the structure of which is disclosed in the prior art *if* there was no obvious way of making it. The prior art must be enabling for it to be novelty destroying. A hypothetical compound that cannot be made is clearly not in the public domain.

Contrast this with genomics. In this field the structure of the compound is generally not novel, even in the patent sense. The gene sequence exists in nature. To qualify as novel the sequence is claimed in forms not found in nature (for example, purified composition, attached to a radioactive label, or attached to an rDNA expression vector). There is no technical issue with the non-obviousness of the sequence's design, as this is not the result of human ingenuity. The technical hurdle in this field is determining (i.e., cloning) the sequence.

All other industrialized countries approach the non-obviousness of novel genes by focusing on the technical hurdle the inventors faced—cloning the gene. For example, the European Patent Office (EPO) in the counterpart application for the *Bell* invention, found that the gene in question was obvious (i.e., lacked inventive step) because it believed there were obvious methods available to clone it.[30] The EPO has also taken a strict stance on the obviousness of recent genomics invention. They recognize that generally there is nothing inventive per se in obtaining such sequence. The current view of the EPO is that a genomics invention will only have an inventive step if the applicant can demonstrate either that obtaining the sequence was in fact a technical achievement or that they have discovered a new or unexpected property associated with the gene. Genomics-based inventions are, therefore, not patented as frequently in foreign patent systems.

Ironically, the European approach is analogous to another U.S. patent law doctrine in synthetic chemistry—the doctrine of structural obviousness or the *Hass-Henze* doctrine.[31] Under this doctrine the courts recognized it was within the state of the art to make certain structural changes to a prior art compound and to expect the new compound to have similar properties. For example, if a prior art herbicide is a large aromatic hydrocarbon having a methyl group at a particular ring position, it would be prima facie obvious to substitute an ethyl for the methyl group and expect the new compound to also be an herbicide. The ethyl compound would only be non-obvious if the inventor could show that the ethyl compound was unexpectedly superior relative to the methyl compound or that the ethyl compound had an unexpected property not shared with the methyl compound.

The *Hass-Henze* doctrine is an example of the non-obviousness standard in practice being tailored to the technical reality of the field of the invention. The European approach to genomics is the same. Just as the *Hass-Henze* doctrine has worked well for the better part of the 20th century, it can be expected that the European approach to genomics will also be successful over time.

---

[30]EPO Technical Board of Appeals Decision No. T0475/93-3.3.4 (1997).

[31]*In re Hass*, 141 F.2d 127 at 130, *available at* 60 U.S.P.Q. 552 (BNA) (CCPA 1944); *In re Henze*, 181 F.2d 196, *available at* 85 U.S.P.Q. 261 (BNA) (CCPA 1950); *In re Papesch*, 315 F.2d 381, *available at* 137 U.S.P.Q. 43 (BNA) (CCPA 1963); *In re Dillon*, 919 F.2d 688, *available at* 16 U.S.P.Q.2d 1897 (BNA) (Fed Cir 1990). See generally, 2-5 Chisum on Patents § 5.04 [b].

Novel gene patents have been limited in scope as a consequence of a heightened disclosure requirement created in a controversial decision by the Federal Circuit in *Regents of the University of California v. Eli Lilly & Co.*[32] By narrowing the scope of some gene patents to the actual sequence disclosed it is possible that *Lilly* might inherently prevent patents on some technologically obvious genes for which *Bell* would otherwise permit a patent. This, however, is not an adequate solution. First, there is still a loss from the public domain of the sequences described in the patent. Second, patent attorneys have adapted their style of drafting to create claims to genera of DNA molecules that meet the *Lilly* standard even though the patent's specification discloses a single DNA molecule.[33] Third, the *Lilly* decision, like the *Bell* and *Deuel* decisions before it, abandoned the "person of ordinary skill in the art" standard for testing adequacy of disclosures in the former case and testing obviousness in the latter. These decisions substitute a rigid rule as to what the ordinarily skilled person is capable of at a particular point in time, thus crippling the patent law's ability to evolve over time with the technology. There is a substantial risk that if this trend continues, patent law will not be able to balance effectively the obviousness and disclosure requirements so that only patents of appropriate scope for non-obvious inventions are granted.

[32]In *Regents of the University of California v. Eli Lilly & Co.*, 119 F.3d 1559, 43 U.S.P.Q.2d (BNA) 1398 (Fed. Cir. 1997), cert. denied, 523 U.S. 1089 (1998), the court held that a patent application containing the sequence of rat insulin cDNA and a protocol on how to clone the closely related human cDNA was invalid to the extent the claims went beyond the rat cDNA because of deficiency in the patent's disclosure. Surprisingly, the protocol to clone the human cDNA was not found to be lacking in enablement and presumptively put the human cDNA into the public domain. The continuing viability of this case has been called into serious question by other members of the Federal Circuit. See, for example, *Enzo Biochem, Inc. v. Gen-Probe, Inc.*, 63 U.S.P.Q.2d 1618 (BNA) (Fed. Cir. 2002)(opinions by Dyke, J.; Rader, J.; and Linn, J.); *Cf. Amgen Inc. v. Hoechst Marion Roussel, Inc.*, 314 F.3d 1313 (Fed. Cir. 2003) (opinion by Michel, J., joined by Schall, J., refusing to extend *Lilly* to biological materials generally).

[33]There are several ways that mere drafting by patent lawyers can meet the *Lilly* standard and recapture scope-of-gene claims without the inventors actually disclosing anything more of substance. It has become accepted practice to define a genus of DNA molecules by a certain percent homology to the disclosed exemplary gene or, even broader, a genus of DNA molecules that encode a protein having a certain percent homology to the protein encoded by the exemplary gene. See, for example, U.S. Pat. No. 6,699,660, "Immediate Early Genes And Methods Of Use Therefore," assigned to Johns Hopkins University School of Medicine (claims to DNA sequences 90 percent identical to disclosed gene sequence or encoding proteins 70 percent identical to disclosed protein sequence). The USPTO also permits patentees to claim a genus of DNA molecules that hybridize to the gene sequence disclosed in the patent. See USPTO Revised Interim Written Description Guidelines Training Materials, Example 9 (http://www.uspto.gov/web/offices/pac/writtendesc.pdf). Ironically, if the University of California's patent had been drafted using either of these approaches that arose in response to *Lilly*, a claim covering a human insulin cDNA would have been patentable even if the technical content of the patent remained unchanged.

The committee therefore recommends that the USPTO[34] and the Federal Circuit abandon the per se rule announced in *Bell* and *Deuel* that prevents the consideration of the technical difficulty faced in obtaining pre-existing genetic sequences and consider adopting an approach similar to other industrialized countries when examining the non-obviousness of gene-sequence-related inventions. For example, the committee believes that the reasonable expectation standard of *In re O'Farrell* is an appropriate test to apply to gene-sequence-related inventions.

INSTITUTE A POSTGRANT OPEN REVIEW PROCEDURE

In the previous chapter we described several grounds for questioning the validity of some proportion of patents being issued in new areas of technology and newly patented technologies. Low or inconsistent patent standards matter for the following reasons:

- In contrast to incentives to genuine innovation, patents on trivial innovations may confer market power or allow firms to use legal resources aggressively as a competitive weapon without consumer benefit.
- Poor patents could encourage more charges of infringement and litigation, raising transaction costs.
- The proliferation of low-quality patents in a technology complicates and raises the cost of licensing or avoiding infringement.
- The uncertainty about the validity of previously issued patents may deter investment in innovation and/or distort its direction.

There are many ways to address patent quality, and elsewhere we consider those related to standards interpretation and the rigor of the USPTO examination process, which is in part a function of resources. Although it is important to conduct reasonably thorough examinations of patent applications, and needed improvements will cost more than we are now spending, given the volume of applications and the fact that only a small percentage of issued patents achieve any commercial importance, there is a point beyond which it is not practical or economical to invest all of the resources that would be needed to ensure uniformly rigorous and timely examination (Lemley, 2001). Nor can the courts be expected to review patents' validity in a timely, efficient manner. Typically, litigation does not occur until 7 to 10 years after a patent is issued and resolution

[34]The USPTO has declined to apply the *Bell* per se rule in at least one instance. *Ex parte Goldgaber*, 41 U.S.P.Q.2d 1172 (BNA) (Bd. Pat. App. & Int.1996). In that case the prior art was a patent that had an actual example of isolating a new protein and a paper example of how to clone it. Goldgaber claimed the gene. The applicant did not appeal this decision to the Federal Circuit, so it is not known whether the USPTO's approach would have been accepted by the court. Nevertheless, it is believed that the USPTO has not been following *Goldgaber* when examining the obviousness of recent gene-related inventions.

is often delayed another 2 to 3 years. The costs of litigation have been documented above.

Another method of improving quality is to weed out invalid patents or revise and narrow the claims of patents by an administrative process after they are issued.[35] Since 1981 the United States has had such processes, known as re-examination, which are available at any time during a patent's period of enforceability. Re-examination has two forms—*ex parte* re-examination initiated by a patent holder, the director of the Patent Office, or a third-party challenger who plays no role in the examination and appeal stages of the proceedings; and *inter partes* re-examination, in which the challenger may participate but until recently has been barred from appealing issues raised and decided in the administrative proceeding.[36] Almost one-half of *ex parte* re-examinations are sought by patent holders hoping to strengthen their patents, usually in the face of newly revealed prior art (Graham et al., 2003). USPTO-initiated re-examinations are very infrequent responses to criticism of issued patents, some of them having been subject to ridicule. Because of the limitations on appeals, *inter partes* re-examinations have also been rare; there were fewer than 25 requests in 2003. Challengers are loathe to forfeit an opportunity to litigate all of the potential validity issues if accused of infringement.

Although that disincentive has been removed,[37] re-examination has another serious drawback. Re-examination may be requested only on the basis of new prior art or prior art considered in the original examination that raises a substantial new question. Issues of patentability, utility, and the adequacy of written description and enablement, which are problematic to varying degrees in different technologies, may not be addressed in re-examinations, only in litigation.

The committee recommends that Congress seriously consider legislation creating an Open Review procedure, enabling third parties to challenge the validity of issued patents on any grounds in an administrative proceeding within the USPTO.[38] It is crucial to its effectiveness that the system provide more timely, lower cost, and more efficient review of granted patents and a wider range of

[35]For reasons of economy and efficiency, the committee rejected expanding pre-patent opposition beyond existing opportunities for third parties to submit prior art. Under pre-grant challenges, if exercised, the cost of patent prosecution and the delays in patent issuance could escalate, whereas the objective of our recommendations is to reduce them. Furthermore, we believe it is more efficient to focus challenges on patent claims as issued rather than claims as originally drafted.

[36]35 U.S.C. § 301-07 (1980) (*ex parte* re-examinations) and 35 U.S.C. § 311-18 (1999) (*Inter partes* re-examination). See Mossinghoff and Kuo (2002) for a discussion of the features and history of these procedures.

[37]Effective for re-examinations begun on or after November 2, 2002, a third-party requester in an *inter partes* re-examination proceeding can appeal to the Federal Circuit a decision by the Board of Patent Appeals and Interferences or to participate in an appeal of a BPAI decision by the patent owner.

[38]A postgrant challenge procedure has been endorsed as part of the USTPO's 21st Century Strategic Plan and recommended by the Federal Trade Commission (2003) in its report on an extensive series of hearings on patents and competition policy.

remedies than the courts are able to provide. If carefully designed and adequately funded, addressing the entire range of patent quality issues, and not compromised by a conflict of interest, the procedure would represent a superior alternative to either re-examination or litigation.

The details of design will determine whether the system is used, whether it is efficient and fair to all parties, and importantly, whether it is subject to abuses that undermine its purpose. Here we recommend some general features (see also Table 4-1). They do not address all of the legislative considerations:

Process

- Any third party requesting a review should bear the burden of persuasion, subject to a preponderance of the evidence standard, that the claims of a patent should be cancelled or amended.
- The Federal District Courts should be able and encouraged to refer issues of patent validity raised in a lawsuit to an Open Review proceeding, confining themselves to resolving issues of infringement. The Department of Justice or the Federal Trade Commission should be able to request the director of the USPTO to initiate a review if they suspect that an invalid patent or patents are being used to adversely affect competition.
- The requesting party would pay a fee, but the challenger and the patent holder would each pay their attorney fees and other costs.
- The challenger would, of course, have access to the history of the patent's prosecution.
- The proceeding would be conducted by an administrative patent judge (APJ) or panel of judges of the U.S. Patent and Trademark Office.
- The APJ would have discretion to allow limited discovery,[39] live testimony of experts, and cross-examination.
- Subject to the Administrative Procedures Act, the USPTO would have broad authority to design procedures drawing on the best practices of other countries but aimed at speed, simplicity, and moderate cost. It should do so in consultation with professionals steeped in the details of the current administrative proceedings—re-examination, re-issues, and interferences—and familiar with their drawbacks.
- In rare cases, circumscribed in regulation, the USPTO should have discretion to continue an Open Review even if the parties decide to settle their disagreement.[40]

[39]A principal source of delays and high costs in litigation, discovery, if permitted, must be carefully circumscribed if the benefits of Open Review are to be realized.

[40]Offsetting the desirability of preventing "collusive" settlements is the need to avoid discouraging potential challengers from using the procedure if they do not have the option of settling a dispute before Open Review has run its course.

TABLE 4-1 Principal Features of U.S. Patent Re-examination, European Opposition, the Proposed U.S. Open Review Procedure, and U.S. Patent Litigation

	SYSTEM				
FEATURES	Original U.S. Re-examination (*ex parte*)	Revised U.S. Re-examination (*inter partes*)	European Opposition System	Committee Recommendation for an Open Review Procedure	U.S. Litigation
Process					
• Timing of initiation	During patent term	During patent term	Within 9 months of issue	Within 12 months of issue[a]	During patent term
• Third-party participation	Initiate and respond to patent owner statement (if filed)	Yes	Yes	Yes	Yes[b]
• Discovery	No[c]	No[c]	No	Discretionary	Yes
• Live testimony	No[d]	No[d]	Yes	Yes	Yes
• Presumption of validity	No[e]	No	Yes	No	Yes
• Who conducts	Examiner (original only if no other option)	Examiner (original only if no other option)	3-member examiner panel (including original)	USPTO administrative law judges	District court judge
• Case referrals by courts and enforcement agencies	No	No	No	Yes	Not applicable

Issues					
• Not patentable subject matter	Only for amended/ new subject matter	Only for amended/ new subject matter	Yes	Yes	Yes
• Novelty or prior art	PA not examined or PA of record if substantial new question	PA not examined or PA of record if substantial new question	Yes	Yes	Yes
• Non-obviousness or inventive step	Yes	Yes	Yes	Yes	Yes
• Utility	Only for amended/ new subject matter	Only for amended/ new subject matter	Yes	Yes	Yes
• Scope of claims	Only for amended/ new subject matter	Only for amended/ new subject matter	Yes	Yes	Yes
• Written description/enablement	Only for amended/ new subject matter	Only for amended/ new subject matter	Yes	Yes	Yes
• Ambiguity of claims	Only for amended/ new subject matter	Only for amended/ new subject matter	Yes	Yes	Yes
Outcomes					
• Confirm, cancel, or amend scope of claims	Yes	Yes	Yes	Yes	Only confirm or cancel
• Appeal to	USPTO administrative law judges and Federal Circuit (for both levels of appeals, patent owner only, not third party)	USPTO administrative law judges (patent owner or third party)	Technical Board of Appeals	Federal Circuit	Federal Circuit
• Third-party future challenge restricted	No third-party challenges; new *ex parte* request must raise new question of patentability	Yes (on any ground third party raised or could be raised)	Only possible in national courts	No	No

continued

TABLE 4-1 Continued

FEATURES	SYSTEM				
	Original U.S. Re-examination (*ex parte*)	Revised U.S. Re-examination (*inter partes*)	European Opposition System	Committee Recommendation for an Open Review Procedure	U.S. Litigation
Duration and Costs					
• Average duration of proceeding	2 years	Insufficient experience	2.1 years (opposition) 2.6 years (appeal)	1 year (objective)	31 months
• Fees	$2520 per request	$8800 per request	€613	Yes (amount not specified)	
• Average or median costs of each party	$10,000-$100,000	Insufficient experience	€30,000-50,000 (both parties)	Not estimated	$1-3 million
• Paid by	Parties	Parties	Parties	Parties	Litigants[f]

[a]Committee majority recommendation; minority supports anytime during patent life.
[b]Third party cannot initiate a challenge on validity or infringement unless there is a reasonable apprehension of suit by the patentee.
[c]Copies of discovery document from litigation may be submitted during the proceeding.
[d]Transcript of testimony from litigation may be submitted during the proceeding.
[e]USPTO Director may initiate based on information provided by enforcement agencies.
[f]Costs may be assessed to the losing party in some cases.

• The review procedure would substitute for *inter partes*- and third-party-initiated *ex parte* re-examination.

Issues and Outcomes

• Validity could be challenged on any ground—that the invention is not patentable subject matter, is not novel, is obvious, lacks utility, or is not properly disclosed.

• Matters previously considered by the patent examiner could be reviewed.

• The outcome would be a confirmation, cancellation, or amendment of the claims in dispute, but claims could not be broadened in a review proceeding.

• Either party could appeal the APJ's decision, first to the Board of Patent Appeals and Interferences, and then to the Court of Appeals for the Federal Circuit. Appeal to the Federal Circuit would invoke estoppel.

The committee is not of one view on the important issue of whether patents should be subject to challenge and review for only a limited time after they are issued, as is the practice in Europe, or for as long as they remain in force. A majority of members recommends that the window for a challenge should be limited to one year from the date of grant so that uncertainty is reduced later in the patent's life. Whenever a patent holder alleges infringement, however, either by filing suit or by notification of an intention to file suit, the review procedure should be available to an accused infringer for a reasonable time. In other words, a review initiated after the one-year window closes would be triggered by an action of the patent holder. A presumption of validity would adhere to a patent after the one-year window closes or to a patent that survives a challenge or is amended in a review proceeding.

A minority of committee members takes strong exception to any time limitation on the exposure of a patent to challenge and review even if the option remains available to an accused infringer. Patents are sometimes issued on speculative utility claims and viewed by others as having no commercial value until another innovator subsequently discovers a valuable utility. By then it might not be possible to challenge the original patent and open the possibility of patenting the valuable utility if the window had closed and the patent owner had not yet attempted to enforce it. Perhaps more frequently, a time limitation would discriminate in favor large companies and institutions with the resources to monitor what patents are being issued. The proponents of having no limitation further point out that although in the EPO opposition system a challenge must be filed within nine months of grant, anyone has standing in most European national patent systems to attempt to invalidate a patent through litigation. In the United States such standing is limited to accused infringers. Obviously, Congress should fully evaluate these opposing positions in considering legislation to create an Open Review system.

In a formal analysis in the accompanying volume of STEP-Board-sponsored research, Levin and Levin (2003) make a strong theoretical case for the welfare gains of adopting an Open Review procedure. These include the prevention of unwarranted monopoly profits, the alignment of patent costs and benefits to genuine novelty and utility, and the reduction in uncertainty for all participants in the relevant market. These benefits depend heavily on two effects or characteristics of the system—first, that it tends to substitute for rather than lead to litigation and, second, that it is less expensive and faster than litigation. It is also conceivable that Open Review, even though it replaced litigation in many cases, could be so popular that its total costs would exceed the costs of litigation, but this outcome is unlikely.[41]

Open Review would be consistent in most basic features with the opposition system available in Europe and other countries (see Table 4-1). In the detailed empirical comparison of re-examination and opposition supported by the STEP Board, Graham and colleagues (2003) find considerable evidence that opposition works reasonably well in many respects. For example, it is used with some frequency. Slightly more than 8 percent of European patents were opposed in the period studied, 1981-1998. Moreover, using citations in other patents as an indicator of value, the opposed patents are more commercially important than the unopposed patents. Finally, the system produces significant changes in outcomes even though the European Patent Office examination process is generally highly regarded for its thoroughness and rigor. Fully one-third of opposed patents are invalidated, and another one-third are amended in the course of opposition. In subsequent research Harhoff (2003) finds evidence from Europe supporting the Levins' prediction that the use of opposition will substitute for subsequent litigation over validity if the process is cheaper, even if it may not be speedier.[42]

Graham and colleagues do, however, confirm testimony of EPO officials to the committee that the agency's opposition process is subject to delays, as long as several years. In fact, the average length of time between patent issuance and the conclusion of opposition is approximately the same as the average time between issuance and the conclusion of litigation in the United States. This appears to be largely a function of the ability of either party in an opposition to extend deadlines for actions indefinitely—a state of affairs that we think should be precluded in the careful design of a U.S. system. Thus, either by legislation or by regulation, Open Review procedures should tightly constrain the schedule to ensure both the timeliness and the lower cost of the process compared with litigation. In particular,

[41]The proportion of patents opposed in Europe is probably higher than it would be in the United States, because the European procedure is the only way to invalidate a patent in all European Patent Convention signatory countries. Enforcement and validity determinations through the courts are on a national basis. Nevertheless, it is impossible to predict usage rates without knowing how attractive the features of Open Review make it relative to litigation and relative to foregoing a challenge altogether.

[42]Again, this experience may not translate to the United States because of system differences.

time limits should be set for discovery and other information-gathering activities and for all responses to actions by the presiding judge or panel. The time limits should not be subject to extension for the convenience of one or both parties but only where meeting the time limit would cause a great hardship or where delay is unavoidable. The objective should be to conclude cases within one year of the request.[43]

It will certainly require additional resources—money, infrastructure, people, and space—to achieve an effectively functioning review procedure in the USPTO; but it should not be assumed that it cannot be done. In fact, it is encouraging that under recent management the Board of Patent Appeals and Interferences has improved the efficiency of its operations, substantially reducing its backlog of cases. We are convinced that when the expanded board is functioning, it will be superior to the district courts in resolving patent validity issues.

In the past, U.S. adoption of a system comparable to opposition has been strongly opposed by some part of the U.S. "independent inventor" community as a potential weapon of large businesses against individuals and small enterprises. Graham and colleagues (2003) show convincingly that this is not so in Europe. Opposed patents are not disproportionately held by small entities nor are large firms disproportionately responsible for initiating oppositions. On the contrary, there is every reason to believe individuals and small businesses would be beneficiaries of an alternative, cheaper, and faster system of resolving patent validity questions. As Lanjouw and Schankermann (2003) show in their chapter of the accompanying volume, it is in litigation that the greater resources of large firms give them substantial advantages both in prosecuting cases to conclusion and in achieving settlements on favorable terms.

We believe that the availability of Open Review, most often occurring within a short time after a patent is issued, will have important side benefits. First, it will encourage firms to review newly issued patents, increasing technology spillovers. Second, it will provide guidance to patent examiners much earlier in the technology cycle than they currently acquire it from court decisions.

STRENGTHEN USPTO CAPABILITIES

The U.S. Patent and Trademark Office is in a double bind. The quality of its output is often questioned and its decisions are widely considered to take too

---

[43]The importance of this objective is underscored by the experience reported to the committee of one firm faced with multiple EPO oppositions that dragged on for a total of eight years without resolution, seriously undermining the firm's ability to enforce its patent rights. If the Federal District Courts were inclined to stay infringement proceedings pending the outcomes of unconstrained USPTO reviews, the ability of patent owners to enforce their rights could be delayed even longer. The USPTO is sensitive to the need to expedite review proceedings, noting in documentation supporting the 21st Century Strategic Plan that electronic filing and processing of cases would contribute to this objective.

much time. The current discussion of the patent fee structure, fee revenue, and USPTO appropriations suggests that many observers believe that the answer lies mainly in providing more financial resources. We believe that more resources are clearly required but that a careful assessment of needs and priorities should precede a determination of how much more. As our committee was not charged nor appropriately constituted to conduct a thorough management review of the USPTO, our list of needs and priorities is not comprehensive nor have we estimated its cost. Nevertheless, we believe the following steps hold the greatest promise for improving performance.

Personnel

In recent years the number of examiners has not kept pace with the increase in the number and complexity of applications, while employee turnover has been rising. Thus, a relatively smaller, more inexperienced workforce is faced with a growing backlog of applications. Congressional appropriators, skeptical of USPTO management, have grown increasingly reluctant to authorize higher personnel ceilings although they have given the office greater flexibility in pay scales to attract new recruits and retain current employees. To relieve their dilemma, administrators resorted to what we considered dubious solutions in the first version of their 21st Century Strategic Plan. First, they proposed to privatize most of the prior art searches by directly contracting or requiring applicants to contract for such services from performers yet to be identified. Second, they proposed to measure application pendency from the initiation of examination following completion of the search process. The plan assumed that the elimination of examiner searches would save on average four hours of examination time, while the new measure of pendency would reduce the average time to disposition from 24 months to 18 months over a period of time.

These proposals generated vigorous criticism at our August 2002 conference and elsewhere. The principal objection was that prior art searching is an integral part of the examination to determine novelty and non-obviousness and that separating the two functions would almost certainly further degrade patent quality. The examiner's need to be familiar with and evaluate the reported prior art would reduce any time saving; hence, the expected reduction in pendency was an arbitrary artifact of redefining the term, not a real gain. As a result of the criticism the USPTO announced that it would conduct a modest-scale, monitored experiment with contracted searches, and it abandoned the redefinition of pendency. We believe that was a wise recourse.

The episode nevertheless underscores the fact that in a patent examination system, as opposed to a patent registration system without any quality control, there is no substitute for having adequate numbers of trained personnel with sufficient time to exercise their considered judgment on the cases assigned to them. Although we do not know how many more examiners are needed to perform quality searches on fewer applications in less time, we are confident that the current number is inadequate to handle the workload now and for some time to come. Other steps may need to be taken to improve examiner competency and reduce turnover.

Electronic Processing

After a series of false starts, the USPTO opted to adopt the European format for storing patent applications in electronic form. The office is working with the EPO to standardize electronic filing in order to increase its usage. This has the advantage of reducing the burden on multinational patent applicants although the format has the distinct drawback of not permitting full-text searches. The USPTO is now implementing an electronic file wrapper and an electronic filing system. The electronic file wrapper allows all those who needed to work on an application anywhere in the USPTO to access the application at their desktops. Among other benefits, this prevents considerable loss of time spent in searching for applications that are moved from one examiner's office to another. From a quality standpoint an electronic file wrapper could include mechanisms for presearching sections of the application against the patent and possibly nonpatent prior art databases using sophisticated language structure searches as opposed to simple word searches. It could also allow for quickly searching claims or phrases in claims against the specification to locate critical parts of the specification without having to read it in its entirety. Another useful capability would be to electronically search the information disclosures submitted by the applicant. In all likelihood other useful capabilities could be developed.

After the 18-month publication of the patent, the electronic file wrapper should be publicly accessible, so that interested parties can follow the examination and better anticipate its results. Even if this might not encourage the submission of patent-defeating or claim-limiting prior art known to competitors, it would inform their decisions about whether to file an Open Review request. It might also be an incentive for the examiner to exercise more care and maintain a more complete record of the examination process. One of the common occurrences in patent prosecution that should be much better documented is in-person or telephone negotiations between examiners and applicants' representatives.

Analytical Capability

The USPTO needs a robust multidisciplinary analytical capability with economic, statistical, management, and program evaluation expertise. The agency currently has a small staff engaged in technology assessment and forecasting. In the past this office performed useful analysis of new technology emergence and patenting rates and their impact on USPTO staffing needs. But now the tasks required are much more substantial and the expertise needed more diverse. The first function of an improved analytical capability is indeed to provide an "early warning system" to help the USPTO anticipate the emergence of new technologies being proposed for patenting. The importance of this capacity is threefold. First, it will help inform decisions to amend the patent classification or create a new technology category or class. Second and closely related, it will inform decisions about hiring or training or reassigning supervisors, examiners, and classifiers. Third, it will help the USPTO anticipate the need to develop or acquire information sources on nonpatent prior art or to hold applications until such resources are obtained and examination issues resolved. If the recent decision of the USPTO leadership to contract out patent application classification in any way undermines the office's ability to detect and respond to the emergence of new technologies, it should be reconsidered.[44]

A second function of improved analytic capability would be to inform management and evaluate proposed administrative changes. The far-reaching reforms proposed in the first 21st Century Strategic Plan were remarkable for their lack of analytical and empirical support. Take as one example the outsourcing of prior art searches. What private resources were already available? What would prices have to be to induce new investment in high-quality search capabilities? How big an increase in costs would there be for different classes of inventors? How would outsourcing affect examiner behavior and time allocation? How would it affect the quality of examination? Answers to these questions were either lacking or guesswork.

The same was true for proposed fee schedule changes, with great potential to affect examination and patent quality either positively or negatively. What would be the impact on application volume of raising fees across the board? Would substitution of an examination fee for the current application fee affect the volume of applications? How would punitive fees on submitting multiple claims in a single application affect drafting? What would be the effect on different technology classes? What are the economics of changing maintenance versus application fees? These critical questions, too, begged for analysis.

---

[44]Some guidance on how to recognize new areas of technology is provided by the Manual of Patent Examining Procedure (MPEP) Chapter 700, section 704.11, stating criteria for requiring submission of information "reasonably necessary to the examination or treatment of a matter in an application."

(a) The examiner's search and preliminary analysis demonstrates that the claimed subject matter cannot be adequately searched by class or keyword among patents and typical sources of nonpatent literature, or (b) either the application file or the lack of relevant prior art found in the examiner's search justifies asking whether the applicant has information that would be relevant to the patentability determination. The first instance generally occurs where the invention as a whole is in a new area of technology that has no patent classification or has a class with few pieces of art that diverge substantially from the nature of the claimed subject matter. In this situation the applicant is likely to be among the most knowledgeable in the art, as evidenced by the scarcity of art, and requiring the applicant's information of areas of search is justified by the need for the applicant's expertise. At a minimum, there needs to be a procedure for aggregating such information from examiners in an art unit or related art units.

A third function of an improved analytical capability would be to support a reliable, consistent, reputable quality assurance process, including determining what the sampling rate and frequency should be across art units and what the evaluation procedure should be. Although such a process should itself be subject to continual improvement, the variability in sample size (and therefore the organizational level at which one could draw a statistically valid conclusion) in recent years should be eliminated. A well-designed quality review process that remains consistent over time is best able to provide useful, credible results to examiners and the public. It can greatly inform if not settle debates about trends in patent quality.

Financial Resources

We cannot precisely estimate the additional resources required to implement our recommendations, but we can order both the budget increases and the budget savings that would be entailed as follows (see Box 4-1).

It is clear that the current budget of the USPTO does not suffice to accomplish these objectives. The patent bar has focused much attention on the fact that for the past several years the fees collected from patent applicants and patent holders have exceeded congressional appropriations to the USPTO by a sub-

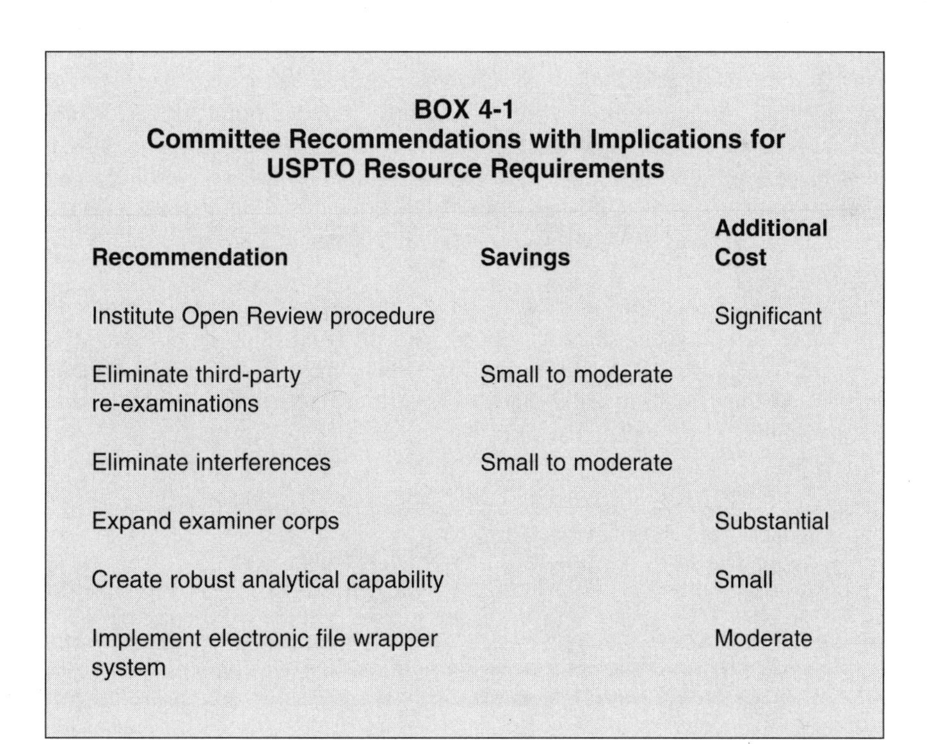

BOX 4-1
Committee Recommendations with Implications for USPTO Resource Requirements

Recommendation	Savings	Additional Cost
Institute Open Review procedure		Significant
Eliminate third-party re-examinations	Small to moderate	
Eliminate interferences	Small to moderate	
Expand examiner corps		Substantial
Create robust analytical capability		Small
Implement electronic file wrapper system		Moderate

stantial margin. Approximately $638 million in revenue over 10 years and an estimated $100 million in fiscal year (FY) 2004 have been spent on other governmental activities.[45] In his FY 2005 budget the President proposes to suspend the practice and devote all of the fees to administration of the office. The corporate patent bar has endorsed an increase in fees provided that none of the revenue is used for other purposes. Certainly these steps would put the USPTO budget closer to what is needed. But it may not be wise to link fee income and expenditures permanently. As a practical matter an abrupt change in the economy could produce a change in revenue but not, given the backlog of applications, a corresponding change in workload. More importantly, Congress should thoroughly consider how fee financing would affect the way the Patent Office conducts its business. Would it, for example, create incentives to issue patents too generously to increase revenue? The patent system serves the broad public purpose of stimulating technological innovation. Its budget should be determined on the basis of what resources are needed to perform the function well.

SHIELD SOME RESEARCH USES OF PATENTED INVENTIONS FROM INFRINGEMENT LIABILITY

In the aftermath of the October 2002 *Madey v. Duke University* decision of the Federal Circuit Court of Appeals, 307 F.3d 1351 (Fed. Cir. 2002), most organized research using patented inventions is subject to demands for licenses and may in some cases be halted by an injunction or assessed money damages for infringement. Although the lawsuit involved the complaint of a former faculty member against a private university employer continuing to use his inventions, a reasonable interpretation of the court's opinion is that formal research enjoys no absolute protection from infringement liability regardless of the institutional venue, the purpose of the inquiry, the origin of the patented inventions, or the use that is made of them.

In the judgment of one member of the Federal Circuit, the *Madey* decision, in combination with a subsequent research exception decision in *Integra Lifesciences I, Ltd. v. Merck KGaA,*[46] cast doubt even on the freedom to investigate into patented inventions to understand them or improve upon them, as had "always been permitted" by the patent system.[47]

[45]Statement of Intellectual Property Owners Association to the House Judiciary Subcommittee on Courts, the Internet and Intellectual Property (April 3, 2003).

[46]*Integra Life Sciences I, Ltd. v. MerckKGaA*, 331 F.3d 860 (Fed. Cir. 2003).

[47]Dissent of Judge Pauline Newman in *Integra Lifesciences.* Indeed, the Federal Circuit on several previous occasions had taken the position that the designing around patents is to be encouraged. See for example, *WMS Gaming Inc. v. Int'l Game Tech.*, 184 F.3d 1339 at 1355 (Fed. Cir. 1999), *Westvaco Corp. v. Int'l Paper Co.*, 991 F.2d 735, 745 (Fed. Cir. 1993), *Yarway Corp. v. Eur-Control USA, Inc.*, 775 F.2d 268 at 277 (Fed. Cir. 1985), *State Indus., Inc. v. A.O. Smith Corp.*, 751 F.2d 1226 at 1235-1236 (Fed. Cir. 1985).

Although the common law "research exception" from liability claimed by Duke University and upheld by the trial court judge but construed very narrowly by the appeals court may never have been robust, it was widely assumed, especially by academic investigators and research administrators, to shield scientific investigation at universities from lawsuits. The STEP-financed survey of scientists and others involved in biotechnology research and development, conducted before the *Madey* decision, suggests that there is widespread indifference to the existence of patents on elements of research, leading to frequent infringement. This may help account for there having been fewer acknowledged research hold-ups or delays and less escalation of research costs attributable to the difficulty of gaining access to patented material than some observers had expected (Walsh et al., 2003). Conceivably, the assumption was also prevalent among patent holders and helps account for their restraint in asserting their rights against research performers.

It is premature to speculate whether the *Madey* decision will result in more frequent patent infringement lawsuits, for example, between patent-holding companies or individuals and universities. There is some evidence that more universities are receiving notices asserting patent rights in 2003 than in 2002.[48] These generally take the form of letters from patent owners' counsel claiming infringement and suggesting or demanding negotiation of licenses or cessation of the activities. Whether these notices are from a few or multiple sources is unclear but perhaps irrelevant. With many more corporations and consultants in the business of asserting patents for royalties, the potential for research disruption and/or cost escalation is present even if the risk of full-blown litigation, injunctions, and damage assessments is not high. The potential disruption or cost will be greater when research institutions are making commercial use of the results of research and/or asserting their patents against commercial enterprises, inviting counter-assertions. Moreover, although some companies and private research sponsors attempt to ensure their freedom of action by examining currently issued patents that may be infringed by a product, service, or research activity, it is much more difficult for university administrations, dealing with large numbers of independent investigators conducting uncoordinated projects supported by multiple sponsors, to exercise such diligence.

The administrative burden on investigators and their institutions and the financial cost of efforts to ensure observance of patent rights could be considerable. At the same time those efforts could be only partially effective because research scientists are often ignorant of the existence of patents. Nevertheless, these are not by themselves compelling reasons to change the outcome of the *Madey* case. Regulatory requirements serving other objectives—for example, financial accounting, human subjects protection, biosafety—are complicated and costly to implement but have been accommodated by the research system.

---

[48]See page 76 above.

We nevertheless believe that there are three other reasons to consider providing some explicit protection from infringement liability. The first has general applicability and relates to Judge Newman's concern that freedom to work on a patented invention is placed in jeopardy by the recent Federal Circuit decisions. That, as she suggests, would represent a fairly radical change in patent law. The other two reasons have to do with the conduct of fundamental research that advances knowledge and provides the building blocks for useful applications. We described and documented these circumstances in the previous chapter. At present they appear largely confined to biotechnology research, but they may extend to other fields where there is a proximity among fundamental and applied research and product development.

- First, with the expansion of patenting of research tools the likelihood that research far removed from commercial applications will entail use of proprietary technology may be increasing.
- Second, at least in biotechnology, restrictions on access to rival-in-use foundational research tools can inhibit realization of their full potential because no single firm can conceive of all of the ways the discovery might be exploited. An example where such restrictions have been imposed, at least for some time, is the BRCA1 gene. Moreover, Henry et al. (2002) find that 68 percent of licenses granted by universities and public labs for genetic inventions were exclusive. The question then becomes what is the net social welfare effect of limiting the number of actors pursuing the development of these technologies.

We believe these circumstances may justify providing some sort of safety valve, but designing a targeted solution is an altogether more difficult matter than deciding whether one is needed. For one thing, not all activities that could be considered research deserve protection. Curiosity-driven inquiry that advances fundamental knowledge perhaps should not be subject to infringement liability, but R&D that is directed at commercializing the patented product should not be free to ignore intellectual property. Where to draw the line is far from obvious. Although much basic research is performed in universities, and companies tend to focus their effects in applied research and development, there is no sharp division of labor, as the Federal Circuit observed in *Madey v. Duke University*, 307 F.3d at 1362 n. 7: "Duke, . . . like other major research institutions of higher learning, is not shy in pursuing an aggressive patent licensing program from which it derives a not insubstantial revenue stream."[49]

Conversely, many corporate laboratories conduct fundamental research whose results are published in the peer-reviewed scientific literature. So if

[49]The benefits and costs of university patenting and licensing stimulated by the Bayh-Dole Act were discussed in an April 2001 STEP workshop held in conjunction with this project. The full proceedings are available on CD-ROM, *Patents in the 21st Century*, obtainable from the STEP Board.

research meriting protection and research not meriting it cannot be clearly distinguished by who performs it or where it takes place, we are left with defining the difference and then trying to apply the definition on a case-by-case basis.

This effort may have been feasible in an earlier era but before the distinctions between basic and applied research or between science and technology broke down. Modern technology is science-intensive and modern science is technology-intensive, with the result that many recent advances are dual in character. Biotechnology inventions, for example, can have immediate applications as diagnostics or therapeutics as well as in research. Mathematical algorithms may function simultaneously as building blocks of knowledge and as bases for commercializable software.

A further complication is that even within the realm of fundamental research there are activities that should not be shielded from liability. An example is the use of research tools whose development depends on the incentive provided by patent protection. How often this is the case is unclear, but however infrequent, here we should encourage, not discourage, the observance of intellectual property to promote investment in the development of new and better research tools. Obtaining licenses to use such technologies may entail an immediate cost in licensing fees and in some cases future costs through reach-through rights, but being denied access to the technologies is not usually a problem because their sole or principal market is research applications.

Other countries have addressed these issues by granting a narrow research exception that hinges on the use of the patented technology rather than the characteristics of the research performer or the intended purpose of the research.[50] Article 27(b) of the European Patent Convention of 1975 as reaffirmed in 1989 declares, "The right conferred by a community patent does not extend to acts done for experimental purposes relating to the subject matter of the patented invention." All European Union members except Austria have incorporated this provision in national law, and in several countries there has been case law interpreting it to mean that researchers of any affiliation may freely use a patented invention to

- determine whether it functions as claimed in the patent;
- determine whether something known to work in certain conditions will work in different conditions;
- discover something unknown about it; or
- improve upon it.

[50]The following description of European law relies on a presentation by Josef Straus, director, Max Planck Institute for Intellectual Property Law, Munich, Germany, to a workshop of the American Association for the Advancement of Science, Washington, D.C., April 24 , 2003 (http://sippi.aaas.org/meetings/04242003/straus_files/frame.htm).

Our search, although not necessarily exhaustive, suggests that other countries that have a statutory research exception have opted for one of comparable (that is, narrow) scope.[51]

In the United States, Congress has adopted research exceptions in three contexts—the first for use of copyrighted material under certain circumstances,[52] the second for "bona fide" research on plant varieties subject to the Plant Variety Protection Act enacted in 1970, and the third in the 1984 Drug Price Competition and Patent Term Restoration (Hatch-Waxman) Act for clinical testing by a generic pharmaceutical company preparing an application to the Food and Drug Administration for marketing approval of a drug subject to an expiring patent. The closest Congress has come to adopting a general research exception was in 1990 when the House Judiciary Committee reported, but neither house passed, a bill that addressed the research tool issue in different terms from, but with similar effect as, the European legislation. That is, it aimed to protect activity intended to gain new knowledge but not the use of patented inventions as research tools.

> It shall not be an act of infringement to make or use a patented invention solely for research or experimentation purposes unless the patented invention has a primary purpose of research or experimentation. If the patented invention has a primary purpose of research or experimentation it shall not be an act of infringement to manufacture or use such invention to study, evaluate, or characterize such invention or to create a product outside the scope of the patent covering such invention.[53]

Other proposals have been advanced, chiefly by legal scholars. Rebecca Eisenberg (1989) has suggested a three-pronged screen. First, no authorization would be needed to use patented inventions for the purpose of checking the validity of the patent holder's claims. This would permit a degree of reverse engineering and have the benefit of reinforcing the peer-review norm of Mertonian science. Second, free use of patented technology would be allowed for the purpose improving upon the invention. In Eisenberg's view such uses would not impinge significantly on the patent holder's financial interest, because the improved product would fall within the scope of the original patent, and its commercialization would therefore require the negotiation of a license and payment

---

[51]These include Canada, Hong Kong, Iceland, Korea, and Japan. Memorandum of Amelia Miazad to Pamela Samuelson, November 27, 2001. The European Patent Convention (EPC) provision would not, in all likelihood, have covered the Duke University experiments, at issue in the *Madey* case, using patented laser equipment as a research instrument, not the object of investigation. It is, however, worth noting that some of the EPC countries have had difficulties making the required distinctions (see United Kingdom Royal Society, 2003). Nonetheless, a proposal along the lines of the EPC has been made for U.S. law (see Strandburg, 2004).

[52]17 U.S.C. § 108(h)(1).

[53]H.R. 5598, the Research, Experimentation, and Competitiveness Act of 1990, introduced by Congressman Robert W. Kastenmeier, 136 Congressional Record, H 7498-99, September 12, 1990.

of royalties. In contrast, research on superseding products would not qualify for the exception because these products would be in competition with the patentee's technology and thus undermine the patent reward. Third, no exception would be available for research use of a patented invention where research is its primary market; otherwise, the financial reward of the patent would be directly undermined.

Similarly, Maureen O'Rourke (2000) has recommended a solution, analogous to the fair use defense of copyright law,[54] requiring an even more discriminating qualitative and quantitative analysis. Under her proposal the court would consider the following five factors:

- the nature of the advance represented by the infringement;
- the purpose of the infringing use;
- the nature and strength of the market failure that prevents a license from being concluded;
- the impact of the use on the patentee's incentives and overall social welfare; and
- the nature of the patented invention.

First, the court would use these factors to determine whether a patented invention could be used without the patent holder's consent. A second inquiry, using the same factors but giving additional weight to the type of market failure, would determine royalties.

An alternative, outlined by Rochelle Dreyfuss (2003), a member of this panel, is to enact legislation exempting a "basic" researcher who demonstrates the noncommercial nature of the researcher's work by agreeing to publish findings promptly and refrain from patenting the discoveries made in the course of using the patented invention. This would be done through the investigator's institution, which would execute a waiver at the outset of the work. Because research is serendipitous and may unexpectedly result in a commercially important discovery requiring patent protection to induce investment in its development, a "buyout" would be permitted to avoid losing these opportunities. The research institution would negotiate directly with the patentee for a license.

Finally, Katherine Strandburg (2004) has recently proposed a categorical statutory exemption for "experimenting on" a patented invention to improve it, whether the experimentation is commercially motivated or not. For experimental use of a patented research tool she proposes some form of compulsory licensing, after a period (perhaps five years) of complete exclusivity, so that the tool inventor is compensated, but others are free to perform tool-based research. She speculates that the delayed compulsory license will rarely be invoked; rather, it will

[54]The copyright fair use defense, 17 U.S.C. § 107, gives those engaged in socially valuable activities, such as news reporting, teaching, scholarship, and research, the right to use copyrighted material without authorization or payment.

serve as an incentive for the negotiation of voluntary licenses during the exclusive period.

The Eisenberg and O'Rourke approaches have several advantages. They are finely tuned to the needs of basic research, while preserving the incentives to innovate in technologies useful in research and elsewhere. They do not discriminate between sectors, for example, between for-profit and nonprofit or university and corporate research performers. And they are broadly consistent with other industrialized countries' policies. It is important to minimize national system differences that may induce or discourage the location of economic activity in one country versus another and that may require eventual negotiation. On the other hand, the distinctions inherent in these proposals are difficult to apply, making the rules less transparent and predictable in application.

Dreyfuss's approach has the advantage of avoiding the need to characterize the invention or the manner of its use or to distinguish between exempt and non-exempt investigators by allowing researchers to self-identify. The government role would be limited to maintaining a registry of waivers. The procedure would be available to scientists in corporate laboratories, although as a practical matter, it is highly unlikely that an industrial employer would allow its corporate R&D staff to commit to publishing all of their results and forego the possibility of patenting or maintaining them as trade secrets. Explicitly, the Dreyfuss proposal is intended to benefit university science and even in some degree to redirect faculty effort away from work with commercial applications or revenue-generating potential.

The assumption is that faculty and university administrators would in appropriate circumstances agree to forego any institutional interest in or financial benefit from the results of the work. That runs counter to research universities' growing investment in technology transfer through patenting and licensing, encouragement of faculty to disclose inventions to central administrations, and aggressive pursuit of industry-sponsored research. Thus, one drawback of her proposal, acknowledged by Dreyfuss, is the friction likely to be generated or exacerbated between university administrators and researchers over when the waiver option should and should not be exercised.

The waiver approach also devalues patents, including patents on research tools, by reducing the size of the market and conceivably leading to the development of products that compete with the patented technology. Dreyfuss's response is that waivers will appeal to the relatively few investigators whose work is truly basic, sharply limiting their impact on patent holders; but in that instance the benefit of shifting effort to more fundamental research and expanding the public domain of research results is also limited. The waiver system is not consistent with the approach taken by other major patenting countries and may require negotiation in the context of patent system harmonization efforts.

The Strandburg proposal suffers from the fact that although it may seem relatively simple to distinguish use of a patented invention to "see how it works" or for the purpose of "improvement" from use of a patented research tool, it may be very difficult in practice. The first two categories should enjoy an absolute exception according to Strandburg, while the latter should not, although it may be subject to compulsory licensing. Is testing a drug against a patented cell receptor "improvement" or "seeing how it works" or is it the use of a tool in precommercial research? If the former, should the drug discoverer be able to file a patent and be exempt from paying royalties? A further drawback is the prevalent hostility in industry and among patent holders generally to any form of compulsory licensing.

The lack of a problem-free formulation does not mean that Congress should not consider the options and try to craft a second-best solution. If it does so, some members of the committee believe that a research exception should be more broadly conditioned than simply requiring a commitment to refrain from patenting the results of the protected research. In this view the conditions should include a showing that the results of the research do not undermine a patentee's commercial markets, a covenant not to use the research results for commercial purposes, and provision for terminating the exemption if the protected research yields patents that are asserted against another party lacking the exemption.

Realistically, the likelihood that Congress will pass research-exception legislation in the absence of compelling circumstances is small. Accordingly, we recommend consideration of administrative action. The federal government could assume liability for patent infringement by investigators whose work it supports under contracts, grants, and cooperative agreements. Under 28 USCA Sec. 1948(a) the federal government can provide "authorization and consent" to contractors who will access U.S. patents in the course of their funded work in the following manner:

> For the purposes of this section, the use or manufacture of an invention described in and covered by a patent of the United States by a contractor, subcontractor, or any person, firm, or corporation for the Government and with the authorization or consent of the Government, shall be construed as use or manufacture for the United States.[55]

The authorization has not often been extended to grantees, although the Department of Energy has exercised this option,[56] and at least one federal district court decision inferred that the government had extended authorization and consent to a research grant recipient accused of infringement.[57]

[55]An alternative legal basis for the government to extend protection to federally supported researchers is the Bayh-Dole Act, under which it may assert a "government use" claim on a patented invention rather than the activity as a whole. This would only shield infringement of a government-supported invention, however, whereas "authorization and consent" could shield the use of any patent.

[56]Communication of Paul Gottlieb, Office of General Counsel, U.S. Department of Energy, (Dec. 16, 2002).

[57]*McMullen Assoc. v. State Board of Higher Education*, 268 F.Supp. 735, 154 U.S.P.Q. 236 (BNA).

This approach, too, has advantages and disadvantages. First, it is somewhat discriminating without relying on nontransparent distinctions that are difficult to apply. It is somewhat targeted at the fundamental research end of the R&D spectrum simply because a large proportion of basic research, as defined in government surveys of R&D performers, is financed by the federal government at universities and other nonprofit institutions. Furthermore, by shifting rather than removing infringement liability, the approach is somewhat sensitive, too, to the rights of research tool as well as other patent holders, who may seek damages from the government in the Court of Federal Claims. The Court of Claims has no injunctive relief authority, but that would rarely be an appropriate remedy in a research infringement case, because from these research uses there would rarely be ongoing commercial losses to the patent holder.

Extension of "authorization and consent" to grantees neither departs from nor advances international patent harmonization efforts, because it does not limit patent rights. Perhaps most importantly, it can be implemented under existing statutory authority, either on the initiative of each federal research agency or on a government-wide basis more uniformly by an amendment to the Office of Management and Budget Circular A-110, which specifies the generic administrative terms of grant relationships between the federal agencies and with nonprofit institutions. Obviously, it should be carefully circumscribed to avoid conferring unrelated legal protections, for example, from tort liability.

A legitimate concern is the impact of such an arrangement on the federal government's liability exposure. Would the government be obliged to defend a large number of cases? Could the government be assessed huge damages? In fact, long-standing Court of Claims case law strictly limits recovery from the federal government for infringement to "reasonable costs and fees." Because the action is equivalent to an eminent domain action and is not a suit in tort, there are no punitive damage awards, let alone treble damages. As the losses attributable to a research infringement are likely to be small, relatively few patent holders are likely to pursue litigation. Of course, they may object that such limitations devalue their patents, undermining incentives to develop improved research tools. That is so, but it is a less severe impact on patentees' financial interests than outright immunity from infringement liability.

Under this proposal, protection would not be available to corporate research, except that conducted under federal contract, nor would it apply to nonfederally supported research at universities. Thus, federally supported research would enjoy a somewhat privileged status. The committee believes that such status is justified by the public interest nature of publicly financed research and that making it relatively more attractive to conduct such research would be beneficial, not a disadvantage. Nevertheless, a certain amount of fundamental research very similar in character to federally funded basic research, for example, supported by philanthropies, would not enjoy protection.

A side benefit of "authorization and consent" is that it would put federally sponsored research in state and private higher education institutions on the same legal footing, without having to undo the Supreme Court's state-sovereign-immunity decisions.

Although our committee is concerned about the discrepancy in status with respect to patent rights introduced by those decisions and we support congressional consideration of a legislative remedy, our committee was not adequately constituted to address either questions of constitutional law or the ramifications of the cases for other areas of intellectual property law.

A more targeted approach to use of the limited immunity of "authorization and consent," outlined by a member of the committee in another context,[58] is to employ it only in cases where access to research tool technologies is not resolved in the marketplace by licensing on reasonable terms. Where use of an important research tool is restricted or prohibitively expensive, an appropriate federal agency such as the National Institutes of Health could award a contract or grant (or conceivably more than one support agreement) incorporating "authorization and consent" for the technology's use. This would be akin to a compulsory license and in all likelihood the threat of its use would lead to a negotiated solution.[59]

On balance the committee recommends that federal research-sponsoring agencies include an explicit "authorization and consent" clause in selected funding instruments as a reasonable step that addresses the need to maintain research tool access, as far as we can ascertain that need so soon after the *Madey* decision.

LIMIT THE SUBJECTIVE ELEMENTS OF PATENT LITIGATION

Among the factors that increase the cost and decrease the predictability of patent infringement litigation are issues unique to U.S. patent jurisprudence that depend on the assessment of a party's state of mind at the time of the alleged infringement or the time of patent application. Accused infringers are usually charged with having done so "willfully," which if proven exposes them to a possible penalty of triple damages. The patent holder frequently is faced with the defenses of "best mode" and "inequitable conduct." The former examines whether the inventor disclosed in an application what the inventory considered to be the best implementation of the invention, while the latter addresses whether the patent attorney intentionally misled the USPTO, usually by failing to disclose important known prior art. Inquiry into these issues requires expensive pretrial discovery. The committee believes that reform in this area would increase predictability of

[58]R. Blackburn. "Owning the Genome?" Presentation to the annual meeting of the Biotechnology Industry Organization, Washington, D.C., June 24, 2003.

[59]Blackburn would require a showing of public harm resulting from the unavailability or high cost of licenses. He does not support general use of "authorization and consent" in federal research grants.

patent dispute outcomes and reduce the cost of litigation without substantially affecting the underlying principles that these aspects of the enforcement system were meant to promote.

"Willful" Infringement and Enhanced Damages[60]

Section 284 of the Patent Act governs damages for patent infringement and provides that in addition to an award "adequate to compensate for the infringement," the court "may increase the damages up to three times." The statute provides no standard for the court to apply in making this determination. In practice the threshold question, usually submitted to a jury, is whether the defendant has been "willful" in the infringement. If the jury finds willfulness, then the judge will determine whether and how much to increase damages within the permitted range based on a list of factors articulated by the Federal Circuit. Exposure to these additional awards substantially raises the stakes for a defendant in patent litigation.

Providing enhanced damages is premised on a principle of deterrence, similar to the rationale for an award of punitive damages in tort litigation. The presumption is that without some substantial additional risk, deliberate infringement becomes more likely, since the potential infringer will ultimately pay the patent holder no more through litigation than through an agreed license. In any event, intentional infringement is viewed as more culpable, justifying punishment. In practice, exposure to a claim of willfulness is not limited to cases of calculated, deliberate infringement. Knowledge of a patent, coupled with a decision to engage in or continue infringing conduct, is enough to trigger the claim. There is no threshold test for having a charge of willfulness considered by the court, so the required level of prefiling investigation by the plaintiff is relatively modest. Therefore, willfulness is asserted in most cases.[61] Because of the stakes for both parties, the issue often overshadows the rest of the litigation.

The most common defense to a claim of willful infringement is good-faith reliance on advice of counsel that the defendant's product or method did not infringe any valid claim. But the shield seems to have morphed into a sword. Courts have held that once a defendant knows of a possible claim of infringement, the defendant is required to obtain an "exculpatory opinion" from an attorney. In the absence of such an opinion the jury may be instructed to infer that any opinion would have been negative. The net result has been a cottage industry

[60]For a general description and background of the doctrine of willful infringement, as well as a discussion of the problems it provokes in patent litigation, see Powers and Carlson (2001). Also see Heffan (1997).

[61]This was not so before the establishment of the Federal Circuit. Then willfulness was pled only when patent infringement was deliberate and even then damages were rarely increased (Heffan, 1997).

of lawyers providing such opinions at a cost ranging from $10,000 to $100,000 per opinion.[62]

Worse, in some business sectors, exposure to claims of willful infringement has led to a practice of deliberately avoiding learning about issued patents, a development sharply at odds with the disclosure function of patent law. Willfulness creates a strong disincentive to read patents, irrespective of whether any infringement allegations are made. The mere existence of the doctrine in its current form means that any time an individual or company learns of a patent that might bear on its products, the company is at risk. Regardless of how the patent comes to light, the company must spend tens of thousands of dollars to obtain an opinion that it is not infringing. And then it foregoes some or possibly all of its attorney-client privilege in the evaluation of the patent.[63] To avoid this situation, in-house counsel and many outside lawyers regularly advise their clients not to read patents if they can avoid it (Lemley and Tangri, 2003; Taylor and Von Tersch, 1998). In this respect patent law stands in contrast with trademark law, which premises willfulness in part on a failure to search for prior marks. Other collateral issues that enter into infringement litigation and raise the complexity and cost of pretrial discovery are the competence of the exculpatory opinion and the propriety of opinion counsel appearing as trial advocates.

Against these costs, complications, and uncertainties there has been no empirical demonstration that the availability of enhanced damages provides substantial additional deterrence over and above that associated with the usual costs and risks of defending an infringement claim, the threat of pretrial injunction relief (rare but potentially devastating to an enterprise), and post-trial award of attorney's fees against deliberate infringement. Thomas Cutter (2004) has analyzed the deterrent effects of enhanced damages for patent infringement generally and concluded that in many circumstances the criteria courts now employ in determining willfulness have an overdeterrent effect—discouraging marginally lawful behavior and taking advantage of the patent disclosure—and therefore undesirable social costs.

Lacking evidence of its beneficial deterrent effect but with evidence of its perverse antidisclosure consequences, the committee recommends elimination of the provision for enhanced damages based on a subjective finding of willful infringement; but we recognize that this is a matter of judgment and that there are a number of alternatives short of elimination that merit consideration.[64] A modest

[62]The process of preparing an opinion is described in Poplawski (2001).

[63]Reliance on the advice of counsel may provide a basis for a successful defense, but by choosing to inject its attorney's advice into the case the defendant waves attorney-client privilege for at least all circumstances and documents relating to that advice and possibly for all advice given before the suit was filed.

[64]As noted above, the Federal Circuit has taken for *en banc* review a willful infringement case, *Knorr-Bremse v. Dana*, 344 F.3d 1336 (Fed. Cir. 2003), and signaled its intention to consider many aspects of the doctrine. Whether the outcome will address the committee's concerns remains to be

step is to abolish the effective requirement that accused infringers obtain and then disclose a written opinion of counsel. Another possibility is to limit inquiry into willful infringement to cases in which the defendant's infringement has already been established. A third alternative that preserves a viable willfulness doctrine but curbs its adverse effects is to require either actual, written notice of infringement from the patentee or deliberate copying of the patentee's invention, knowing it to be patented, as a predicate for willful infringement (Federal Trade Commission, 2003; Lemley and Tangri, 2003). If some form of willfulness doctrine is retained, there is the question by how much should damages be enhanced. One answer is by the least amount needed to deter deliberate copying and make the victims whole. Lemley and Tangri suggest that in most instances awarding successful plaintiffs their attorney fees will suffice as an adequate penalty. Finally, modification or elimination of willful infringement raises questions about the status of the "duty of care" to avoid patent infringement. This is a matter we did not address that merits further consideration.

"Best Mode" Defense[65]

Section 112 of the Patent Act requires that an application "set forth the best mode contemplated by the inventor of carrying out his invention." As interpreted by the Federal Circuit, this requirement is judged by a two-part test; first, did the inventor, at the time of filing, know of a mode of practicing the invention that the inventor believed was preferable to others; and second, was the best mode adequately disclosed, in light of the scope of the claimed invention and the level of skill in the art. The first test is inherently subjective, focusing on the inventor's state of mind; and although the second test is objective, it is not precise.

The best-mode standard is different from Section 112's "enablement" requirement, which goes to the sufficiency of the disclosure to teach one of ordinary skill to implement the invention. Best mode requires in addition that if the inventor knows of particular materials or processes for implementing a claimed invention that the inventor believes are most effective, they must be revealed. For example, consider a claim that states a range of temperatures for operation of a method. If at the time of filing the application the inventor believed that a particular temperature or narrower range was optimal, failure to disclose it—even if unintentional—may result in a finding that the claim is invalid. As with other invalidity defenses, establishing a best-mode violation requires "clear and convincing" proof; and the defense is applied only on a claim-by-claim basis. However, if

seen. The case raises the question of whether willful infringement should be presumed when an infringer failed to obtain an opinion of counsel before infringing or invokes attorney-client privilege or the work product doctrine to avoid disclosing an opinion obtained.

[65]For a general description and background of the best mode defense, see Chisum (1997) and Hofer and Fitzgerald (1995).

intent to deceive is shown, the same proof can establish "inequitable conduct," and the entire patent may be unenforceable.

Only the United States imposes a best-mode requirement. Its goal is to motivate more extensive disclosure to the public by increasing the risk of withholding related information as a trade secret. As explained by the Federal Circuit's predecessor court, the purpose of the requirement is to "restrain inventors from applying for patents while at the same time concealing from the public preferred embodiments of their invention which they have in fact conceived."[66]

Analysis of the best-mode defense is made as of the time the inventor filed the original application; there is no obligation to "update" the application with information discovered during prosecution of the patent. Moreover, the defense applies only to information and belief personal to the inventor, and cannot be established by imputation of knowledge of others in the inventor's company or working group. Therefore, this doctrine as applied gives only limited assurance that the best mode will be disclosed.

Because the defense depends on historical facts and because the inventor's state of mind usually can be established only by circumstantial evidence, litigation over this issue—especially pretrial discovery—can be extensive and time-consuming. Foreign patent applicants also criticize the doctrine as unfair, since their previously filed foreign applications cannot simply be translated for filing in the United States without attending to this unusual additional requirement.

Given the cost and inefficiency of this defense, its limited contribution to the inventor's motivation to disclose beyond that already provided by the enablement provisions of Section 112, its dependence on a system of pretrial discovery, and its inconsistencies with European and Japanese patent laws, the committee recommends that the best-mode requirement be eliminated.

Inequitable Conduct Defense[67]

Even when a patent claim is valid, if obtained through fraud, it is deemed unenforceable. This concept is codified in Section 282 of the Patent Act. The defense of inequitable conduct applies when the patent applicant has made a material misstatement or omission with intent to deceive the USPTO. Like invalidity, unenforceability through inequitable conduct must be proved through "clear and convincing" evidence. However, unlike questions of invalidity, inequitable conduct is decided by a judge, not a jury.

Inequitable conduct requires proof of both materiality of the information and intent to deceive. Materiality has been measured by a standard similar to that applied in cases of securities fraud: whether there is a substantial likelihood that a

[66]*In re Gay*, 309 F.2d 769 at 772 (C.C.P.A. 1962).

[67]For a general description and background on inequitable conduct, see Chisum (1997).

"reasonable examiner" would have considered the information important in deciding whether to issue the patent. A 1992 USPTO rule change appears to have raised the bar on materiality so that now, arguably, a defendant must prove that the information, properly disclosed, would have led the USPTO to reject the relevant claim.[68] Intent is, of course, a subjective issue, directed at the state of mind of the patent applicant or the applicant's attorney. Once a judge has determined that a threshold level of materiality and intent have been proved, the court will consider all of the circumstances surrounding the applicant's conduct and balance the level of materiality and intent to determine if inequitable conduct exists. In other words, a high level of one might offset a low level of the other so that in some cases inequitable conduct is found despite very little evidence of deliberate misconduct.

Examples of behavior punished as inequitable conduct include failure to cite a known prior art reference unless it is merely "cumulative" to those already cited, "burying" a material reference in a stack of irrelevant information, submitting false or misleading declarations related to dates of invention or enablement, and failure to disclose offers for sale and public uses that would make the claim invalid under Section 102(b) of the Patent Act. The consequences of a finding of inequitable conduct can be severe. First, the entire patent, not just the relevant claim or claims, is rendered unenforceable. Second, other patents in the same family may be deemed "infected" by the fraud. Third, the defendant may be awarded attorneys fees under the "exceptional case" standard of Section 285 of the Patent Act. Finally, the patentee may be exposed to an antitrust claim.[69] As with willful infringement and the best-mode defense, discovery is more complex and expensive. Moreover, because the level of disclosure to the USPTO usually involves choices made by the patent attorney, issues of the scope and waiver of attorney-client privilege are implicated. Another major complaint is that the defense is asserted too freely. One judicial opinion commented: "[T]he habit of charging inequitable conduct in almost every major patent case has become an absolute plague."[70]

If invalidity, disciplinary action, and reputational concerns are not sufficient deterrent to misconduct, other civil and even criminal remedies exist—antitrust, unfair competition, common law fraud, and tortuous interference. Moreover, since the creation of the inequitable conduct doctrine by the courts, other safeguards

---

[68]The effect of the USPTO rule change has not yet been decided by the Federal Circuit. In *Molins PLC v. Textron.*, 48 F.3d 1172 at 1179 n.8 (Fed. Cir. 1995), the court merely observed that because administrative rules are not retroactive, and the case arose before the rule change, it did not have to address the issue. However, in *Semiconductor Energy Laboratory Co., Ltd. v. Samsung Elecs Co.*, 204 F.3d 1368 at 1374 (Fed. Cir. 2000), the court applied the new rule to a patent issuing on an application filed after 1992, without any discussion of whether the old standard should apply.

[69]See *Walker Process Equipment, Inc. v. Food Machinery & Chemical Corp.*, 382 U.S. 172 (1965).

[70]*Burlington Industries, v. Dayco Corp.*, 849 F.2d 1418 at 1422 (Fed. Cir. 1988).

have been adopted by Congress and the USPTO to support the integrity of the patent system. These include third-party- and USPTO-initiated re-examination on withheld prior art, publication of pending applications, and third-party access to pending prosecution papers and the ability to submit material information. The Open Review process we propose would also contribute to the integrity of the system.

In view of its cost and limited deterrent value the committee recommends the elimination of the inequitable conduct doctrine or changes in its implementation. The latter might include ending the inference of intent from the materiality of the information that was withheld, de novo review by the Federal Circuit of district court findings of inequitable conduct, award of attorney's fees to a prevailing patentee, or referral to the USPTO for re-examination and disciplinary action. Any of these changes would have the effect of discouraging resort to the inequitable conduct defense and therefore reducing its cost.

Our recommendations would almost certainly simplify litigation and curb unproductive discovery and thereby reduce its expense, but to what extent? And if only one or two rather than all three elements of patent litigation areas of law were reformed, which would yield the largest litigation cost saving? We are not certain, but we have benefited from the opinions of a large group of highly experienced patent litigators, most of them in private practice. Knowing the committee's interest in these questions, the fellows (senior members and former officers) of the American Intellectual Property Law Association conducted an informal survey of their colleagues on the three elements of litigation considered. A substantial minority of 93 respondents supported the modification or elimination of one or more of these rules even though they are beneficiaries of the complexity and cost of patent litigation. A slight majority considered them significant cost drivers, although some of them suggested that uncontrolled discovery was a problem of civil litigation generally. Respondents split evenly in identifying willful infringement versus inequitable conduct as the main cost factor. Best mode ranked a distant third.

HARMONIZE THE U.S., EUROPEAN, AND JAPANESE PATENT EXAMINATION SYSTEMS

As early as 1966 a presidential commission appointed by President Johnson recommended that the United States adopt the otherwise universal first-inventor-to-file basis for determining patent priority as a step toward making the major industrial countries' patent systems compatible. The Carter administration's policy review on innovation and Commerce Secretary Mosbacher's commission on the patent system in the first Bush administration also urged progress toward harmonization. The adoption of the 20-years-from-filing-patent term and publication of most patent applications at 18 months, both under the TRIPS agreement, were important recent steps in that direction. The United States is currently engaged in negotiations under the World Intellectual Property Organization (WIPO) aimed at substantive patent law harmonization. These developments are a recognition that in an increasingly integrated global economy, differences in patent law create redundancy and inconsistencies that raise the cost of doing international business.

The committee did not consider the thorny issues associated with reconciling differences in intellectual property protection between developing and industrialized countries but is primarily concerned with differences in patent examination among the latter, especially the United States, Europe, and Japan. In that context greater harmonization has taken on more urgency with the increase in patent filings generally and the increase in multinational filings in particular (see Figure 4-1).

Each of the three major patent offices has had difficulty coping with the surge in applications, yet work sharing is minimal. For most commercially important inventions, technically and legally skilled patent examiners in each office analyze the same application, search more or less the same prior art, and perform similar examinations, sometimes with identical results, sometimes with results dictated by differences in law.

The committee believes that the United States, Europe, Japan, and other countries should continue to harmonize substantive laws regarding patentability, application priority, rules of prior art, and standards of examination with the objective of establishing systems of reciprocity or mutual recognition of the results of searches and examinations. This goal will require changes in law and practice on all sides. The committee members agree that the following are among the principal differences that need to be reconciled, and we agree on the preferred terms of an agreement on patent system harmonization.

First-to-Invent Versus First-Inventor-to-File Priority

The United States should conform its law to that of every other country and accept the first-inventor-to-file system. There are several reasons for this shift. First, the discrepancy means not only that in some cases different people will own patents on the same invention in different countries but also that there are radical differences in procedure. The United States has an elaborate legal mechanism, both in the USPTO and in the courts, for determining who was the first to invent. Because the rest of the world has no analogous process, foreign patent applicants are subject to uncertainty and perhaps challenges that are entirely unfamiliar. The governments tend to view U.S. acquiescence to the first-to-file as the cornerstone of international harmonization.

Second, U.S. inventors also file their applications in ignorance of whether they are the first or second to invent and when an opponent might be expected to file. For those subject to challenge under first-to-invent, the proceeding is costly and often very protracted; frequently it moves from a USPTO administrative pro-

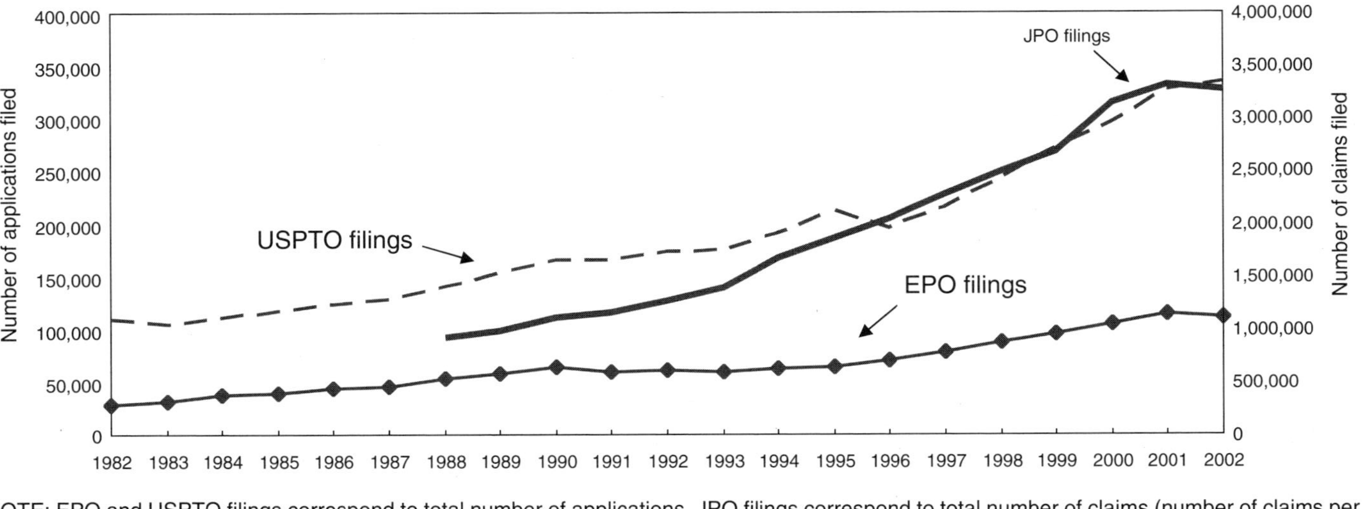

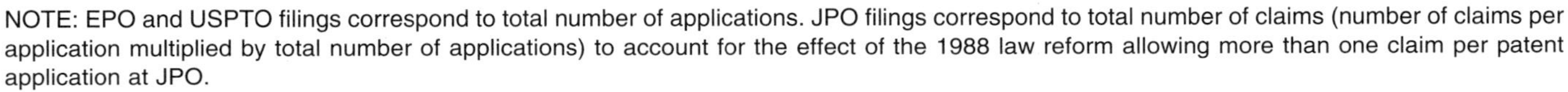
NOTE: EPO and USPTO filings correspond to total number of applications. JPO filings correspond to total number of claims (number of claims per application multiplied by total number of applications) to account for the effect of the 1988 law reform allowing more than one claim per patent application at JPO.

FIGURE 4-1 USPTO, EPO, and JPO patent application filings, 1982-2002. SOURCE: OECD (2003).

ceeding to full court litigation. In both venues it is not only evidence of who first reduced the invention to practice that is at issue but also questions of proof of conception, diligence, abandonment, suppression, and concealment, some of them requiring inquiry into what an inventor thought and when the inventor thought it.

A third reason to adopt the first-inventor-to-file priority basis is that for the overwhelming majority of applicants, that is the system the United States has. Of the more than 300,000 applications the USPTO receives each year only about 200 to 250—less than 0.1 percent—end up in interference proceedings because a second filer claims to be the first inventor.

There are, nonetheless, three concerns that merit attention in considering whether to abandon first-to-invent. The first concern is how often first inventors would be unfairly deprived of their inventions by second inventors who happened to file with the Patent Office first? The answer, it turns out, is a nontrivial number or at least a non-negligible proportion of applicants involved in interferences. Lemley and Chien (2003) examined two sets of interference cases—first, 76 final adjudications by the Board of Patent Appeals and Interferences (BPAI) between 1990 and 1991 that were decided by determining who was the first inventor; and second, a random selection of the few hundred interference proceedings reported on the BPAI web site between 1997 and 2003. They concluded that second filers won approximately 43 percent of the cases. Nevertheless, in a large proportion of these cases first- and second-filers' invention dates were so close as to be nearly simultaneous.

A second concern is the inducement inherent in a first-to-file system to file early and perhaps before the invention is fully characterized, which could be a source of patent quality deterioration. The incentive for early filing surely exists but is mitigated by two factors. First is provisional application filing whereby inventors who file a complete technical disclosure secure priority rights without a major expenditure of resources for legal services. This allows the applicant a year to characterize, refine, consider claims for, and assess the commercial value of an invention before submitting a formal application. The second mitigating factor is that inventors already have significant incentives to file applications early, for instance any inventor who seeks protection outside the United States competes in a first-to-file system.

Most important from a fairness and a political point of view, first-to-file is claimed to disadvantage individual inventors and small business, who may not have the resources to be as fast as large companies. This has been the premise of very effective "independent inventor" opposition to first-to-file and harmonization generally for a very long time. To illuminate the issue Gerald Mossinghoff (2002) studied all 2,848 interference decisions between 1983 and 2000 to determine whether small inventors were more likely to prevail in priority disputes. He found that the first-to-invent system did not benefit small inventors on average. Of that number, 203 were decided in favor of a small entity filing second, but in 201 other cases small-entity first filers lost. For another perspective on the small-entity issue, Lemley and Chien (2003) examined on whose behalf the interference cases in their study were initiated. Strikingly, of the 94 initiators for which status data were available, 77 percent were large firms while only 18 percent were small entities. Of the responding parties 43 percent were individuals or small businesses while 53 percent were large entities. Both sets of evidence support the conclusion that the first-to-invent system is not working to the benefit of small entities; rather, in the preponderance of cases large firms are ensnaring small companies in complex, costly interference proceedings. Even if this were not so, small businesses increasingly oriented toward international markets might prefer harmonization as a way to reduce the total costs of multimarket protection. According to a 2002 General Accounting Office survey, 70 percent of small business respondents agreed with that objective.

Grace Period

The United States should retain and seek to persuade other countries to adopt a grace period, allowing someone to file a patent application within one year of publication of its details without having the publication considered prior art precluding a patent grant.[71] This provision encourages early disclosure and is especially beneficial for dissemination of academic research results that may have commercial application. As other countries try to accelerate the transfer of technology from public research organizations to private firms through patents and licensing, the idea of a grace period is likely to become more widely accepted. Germany recently adopted such a provision.

Best-Mode Requirement

The "best mode" requirement, having no analog in foreign patent law, imposes an additional burden and element of uncertainty on foreign patentees in the United States. This, in addition to its dependence on discovery aimed at uncovering inventor records and intentions, justifies its removal from U.S. patent law.

Prior Art

In the interest of arriving at a uniform definition of prior art, the United States should remove its limitation on non-published prior art and its rule that foreign patents and patent applications may not be recognized as prior art as of their filing dates. In connection with moving to a first-inventor-to-file system, the foreign patent prior art rule for unpublished prior patent applications should also

[71] 35 U.S.C. § 102(b).

be adopted. A common misconception about the EPO and other foreign systems like that of the EPO is that they are winner-take-all systems similar to the U.S. interference proceeding. A difference in prior art treatment, however, prevents this from occurring. Abroad an unpublished prior patent application is available for prior art purposes only under the novelty standard. It cannot be used in a non-obviousness (or equivalent) rejection. This allows the later filing applicant to obtain claims to a disclosed aspect of the invention that is novel with respect to the prior application even if it would have been obvious. This has the affect of giving some reward to near simultaneous inventors. Where the second to file is first with a commercially important embodiment of the invention, the foreign rule increases cross-licensing and enhances competition in the marketplace.

Application Publication

The United States should abandon its exception to the rule of publication after 18 months for applicants not intending to patent abroad. This, too, would promote the disclosure purpose of the patent system. Eliminating the non-publication option would minimize the uncertainty associated with submarine patents, which remain a problem as a consequence of the continuation practice, enabling an applicant to abandon one application and file a continuation or pursue an application to issue while maintaining a continuation on file—in either case in the hope of winning a better patent eventually. Moreover, universal publication would extend to all patentees the provisional rights under 35 U.S.C. Sec. 154(d) (2000) that give a patentee a reasonable royalty for infringement that occurs after publication but before patent issuance under certain conditions (Lemley and Moore, 2004).

Other Issues

There are other differences regarding the scope of patentable subject matter and the standards for non-obviousness and utility that we have not examined in detail and for which we therefore have no precise prescriptions. Given that patent laws are part of historically evolved national legal systems there may be limits to harmonization, but these are likely to recede over time as the international economy becomes more integrated and enterprises more dependent on global markets. Reconciling patent system differences will be challenging but would make the outcome of this so-called "deep harmonization" more rewarding.

The committee supports the pursuit of harmonization through WIPO but recognizes the difficulty of achieving agreement among 180-odd countries with widely divergent views of intellectual property protection generally and the patent system in particular. There is a risk that harmonization of the three major patent systems could be sidetracked by disagreements between the developing and developed countries. We believe that harmonization should and can be pursued in trilateral or even bilateral negotiations or on selected issues whose results, if there is agreement, will have a beneficial demonstration effect on other countries. In the meantime, the practice of convening international panels from the three patent offices to explore common approaches to search and examinations in new technological areas should be continued. This practice is helpful not only in identifying issues for negotiation but, more immediately, in informing patent applicants how their inventions are likely to be treated in each of the patent offices.

The committee recognizes that its proposals, apart from foreign adoption of a grace period, would represent U.S. conformity with other patent systems and may be subject to the charge that we favor "Europeanizing" the U.S. patent system. That is a narrow view. It presumes that only the items enumerated are part of a negotiated package. It implies that the U.S. system features we propose changing are important to its integrity. We disagree. Most important, it ignores what we expect to be the benefits of harmonized priority and examination procedures for U.S. inventors, whether large or small entities—first, faster, more predictable determinations of patentability; second, simplified, less costly litigation; and third, less redundancy and much lower costs in establishing global patent protection.

References

Advisory Commission on Patent Law Reform. (1992). *Report of the Advisory Commission on Patent Law Reform*. U.S. Department of Commerce.

Advisory Committee on Industrial Innovation. (1978). *Final Reports: Patent Policy Report*. U.S. Department of Commerce.

Allison, J. and M. Lemley. (1998). "Empirical Evidence on the Validity of Litigated Patents." *AIPLA Quarterly Journal* 26:185-277.

Allison, J. and M. Lemley. (2000). "How Federal Circuit Judges Vote in Patent Validity Cases." *Florida State University Law Review* 27(3):745-766.

Allison. J. and M. Lemley. (2002). "The Growing Complexity of the United States Patent System." *Boston University Law Review* 82(1):77-144.

Allison, J., and E. Tiller. (2003). "Internet Business Method Patents." In *Patents in the Knowledge-Based Economy*, W. Cohen and S. Merrill, eds. Washington, D.C.: The National Academies Press.

American Intellectual Property Law Association. (2003). *AIPLA Report of Economic Survey, 2003*. Washington, D.C.: Fetzer-Kraus, Inc.

Anand, B. and T. Khanna. (2000). "The Structure of Licensing Contracts." *Journal of Industrial Economics* 48(1):103-135.

Arora, A., A. Fosfuri, and A. Gambardella. (2001). *Markets for Technology: Economics of Innovation and Corporate Strategy*. Cambridge, Mass.: MIT Press.

Arora, A., M. Ceccagnoli, and W. Cohen. (2002). "R&D and the Patent Premium." Unpublished Paper (Feb.).

Arundel, A., J. Cobbenhagen, and N. Schall. (2002). *The Acquisition and Protection of Competencies by Enterprises*. Report for EIMS project 98/180, DG Enterprise, European Union, Luxembourg.

Association of University Technology Managers. (2003). *AUTM Licensing Survey: FY 2000*. Northbrook, IL: Association of University Technology Managers.

Baldwin J., P. Hanel, and D. Sabourin. (2000). "Determinants of Innovative Activity in Canadian Manufacturing Firms: The Role of Intellectual Property Rights." Statistics Canada Working Paper No. 122 (Mar.). Available at http://www.statcan.ca/cgi-bin/downpub/listpub.cgi?catno=11F0019MIE2000122.

Barr, R. (2002). "Statement to the Joint FTC/DOJ Hearing on Competition and Intellectual Property Law and Policy in the Knowledge-Based Economy, February 28, 2002." Available at http://www.ftc.gov/opp/intellect/index.htm.

Bar-Shalom, A. and R. Cook-Deegan. (2002). "Patents and Innovation in Cancer Therapeutics: Lessons from CellPro." *Millbank Quarterly* 80(4):637-676.

Barton, J. (2000). "Reforming the Patent System." *Science* 287(5460):1933-1934.

Barton, J. (2002). "Antitrust Treatment of Oligopolies with Mutually Blocking Patent Portfolios." *Antitrust Law Journal* 69(3):851-882.

Barton, J. (2003). "Non-obviousness." *IDEA: The Journal of Law and Technology* 43:475.

Bekkers, R., G. Duysters, and B. Verspagen. (2002) "IPR, strategic technology agreements and market structure: The case of GSM." *Research Policy* 31:1141-1161.

Berman, B. (2002). "From Tech Transfer to Joint Ventures – Part 1." *Cafezine*. Available at http://www.cafezine.com/index_article.asp?deptId=5&id=555&page=1.

Bernstein, M. (1955). *Regulating Business by Independent Commission*. Princeton, N.J.: Princeton University Press.

Bessen, J. and E. Maskin. (2000). "Sequential Innovation, Patents and Imitation." MIT Department of Economics Working Paper Number 00-01. Available at http://papers.ssrn.com/paper.taf?abstract_id=206189.

Blanton, K. (2002). "Corporate Takeover." *Boston Globe* (24 Feb. 2002): Magazine.

Branstetter, L., R. Fisman, and C. Foley. (2003). "Do Stronger Intellectual Property Rights Increase International Technology Transfer? Empirical Evidence from U.S. Firm-Level Panel Data." Unpublished Paper. Available at http://econ.worldbank.org/files/35589_wps3305.pdf.

Burk, D. and M. Lemley. (2003a). "Policy Levers in Patent Law." *Virginia Law Review* 89(7):1575-1696.

Burk, D. and M. Lemley. (2003b). "Biotechnology's Uncertainty Principle." University of Minnesota School of Law Public Law and Theory Research Paper No. 03-5.

Bush, V. (1945). *Science, the Endless Frontier*. Washington, D.C.: National Science Foundation.

Chisum, D. (1997). "Best Mode Concealment and Inequitable Conduct In Patent Procurement: A Nutshell, A Review of Recent Federal Circuit Cases and A Plea for Modest Reform." *Santa Clara Computer and High Technology Law Journal* 13:277-319.

Cho, M., S. Illangasekare, M. Weaver, D.G.B. Leonard, and J.F. Merz. (2003). "Effects of Patents and Licenses on the Provision of Clinical Genetic Testing Services." *Journal of Molecular Diagnostics* 5(1):3-8.

Clarke, R. (2003). "U.S. Continuity Law and Its Impact on the Comparative Patenting Rates of the US, Japan, and European Patent Offices." *Journal of the Patent and Trademark Office Society* 8(Apr.):335-349.

Cockburn, I. (2004). "The Changing Structure of the Pharmaceutical Industry." *Health Affairs* 23(1):10-22.

Cockburn, I., S. Kortum, and S. Stern. (2003). "Are All Patent Examiners Equal? Examiners, Patent Characteristics, and Litigation Outcomes." In *Patents in the Knowledge-Based Economy*, W. Cohen and S. Merrill, eds. Washington, D.C.: The National Academies Press.

Cohen, J. and M. Lemley (2001). "Patent Scope and Innovation in the Software Industry." *California Law Review*. 89:1-58.

Cohen, W. (2002). "The Pro-Patent Movement in the United States: Indicators and Impacts." Unpublished Paper.

Cohen, W., A. Goto, A. Nagata, R. Nelson, and J. Walsh. (2002). "R&D Spillovers, Patents and the Incentives to Innovate in Japan and the United States." *Research Policy* 31:1349-1367.

Cohen, W., R. Nelson, and J. Walsh. (2000). "Protecting Their Intellectual Assets: Appropriability Conditions and Why U.S. Manufacturing Firms Patent (or Not)." NBER Working Paper 7552. Available at http://www.nber.org/papers/w7552.

Cohen, W. and S. Merrill. (2003). "Introduction." In *Patents in the Knowledge-Based Economy,* W. Cohen and S. Merrill, eds. Washington, D.C.: The National Academies Press.

Commission of the European Communities. (2000). *Proposal for a Council Regulation on the Community Patent* (Aug. 1). Available at http://europa.eu.int/comm/internal_market/en/indprop/patent/412en.pdf.

Cutter, T. (2004). "An Economic Analysis of Enhanced Damages and Attorneys' Fees for Willful Infringement." Paper presented to the Symposium on Willful Infringement sponsored by the Oracle Corporation and the George Washington University Law School, Washington, D.C., Mar. 19.

Desmond, R. (1993). "Nothing Seems Obvious to the Court of Appeals for the Federal Circuit: The Federal Circuit, Unchecked by the Supreme Court, Transforms the Standard of Obviousness Under the Patent Law." *Loyola of Los Angeles Law Review* 26:455-490.

Dolak, L. (2000). "As if You Didn't Have Enough to Worry About: Current Ethics Issues for Intellectual Property Practitioners." *Journal of the Patent and Trademark Office Society* 82:235.

Dreyfuss, R. (1989). "The Federal Circuit: A Case Study in Specialized Courts." *New York University Law Review* 64:1.

Dreyfuss, R. (2003). "Varying the Course." In *Perspectives on Properties of the Human Genome Project*, F. Kieff, ed. New York, NY: Academic Press.

Dreyfuss, R. (forthcoming). "The Federal Circuit: A Continuing Experiment in Specialization." *Case Western Reserve Law Journal.*

Dunner, D., J. Jakes, and J. Karceski. (1995). "A Statistical Look at the Federal Circuit's Patent Decisions: 1982-1994." *Federal Circuit Bar Journal* 5:151-180.

Eisenberg, R. (1989). "Patents and the Progress of Science: Exclusive Rights and Experimental Use." *University of Chicago Law Review* 56:1017-1086.

European Patent Office. (1999). Case Law of the Boards of Appeal. Available at http://www.european-patent-office.org/legal/case_law/e/index.htm.

Federal Trade Commission. (2003). *To Promote Innovation: The Proper Balance of Competition and Patent Law and Policy*. Available at http://www.ftc.gov/os/2003/10/innovationrpt.pdf.

Federico, P. (1956). "Adjudicated Patent Statistics." *Journal of the Patent Office Society* 38(4):233-249.

Gallini, N.T. (1984). "Deterrence by Market Sharing: A Strategic Incentive for Licensing." *American Economic Review* 74(5):931-941.

Gallini, N. (1992). "Patent Policy and Costly Imitation." *Rand Journal of Economics* 23:52-63.

Gallini, N. (2001). "How Well Is the U.S. Patent System Working." Unpublished Paper (Jul.).

Gallini, N. (2002). "The Economics of Patents: Lessons from Recent U.S. Patent Reform." *Journal of Economic Perspectives* 16(2):131-154.

Graham, S., B. Hall, D. Harhoff, and D. Mowery. (2003). "Patent Quality Control: A Comparison of U.S. Patent Re-examinations and European Patent Oppositions." In *Patents in the Knowledge-Based Economy*, W. Cohen and S. Merrill, eds. Washington, D.C.: The National Academies Press.

Graham, S. and D. Mowery. (2003). "Intellectual Property Protection in the U.S. Software Industry." In *Patents in the Knowledge-Based Economy,* W. Cohen and S. Merrill, eds. Washington, D.C.: The National Academies Press.

Green, J. and S. Scotchmer. (1995). "On the Division of Profit in Sequential Innovation." *RAND Journal of Economics* 26(1):20-33.

Hall, B. (2003a). "Exploring the Patent Explosion." Unpublished paper. Available at http://emlab.berkeley.edu/users/bhhall/papers/BHH%20Mannheim03.pdf.

Hall, B. (2003b). "Business Method Patents, Innovation, and Policy." Unpublished Paper. Available at http://emlab.berkeley.edu/users/bhhall/papers/BHH%20on%20BMP%20May03WP.pdf.

Hall, B. and R. Ziedonis. (2001). "The Patent Paradox Revisited: An Empirical Study of Patenting in the U.S. Semiconductor Industry." *RAND Journal of Economics* 32(1):101-128.

Harhoff, D. (2003). "Legal Challenges to Patent Validity in the U.S. and Europe." Presentation to OECD Conference on IPR, Innovation, and Economic Performance (August 28, 2003). Available at http://www.oecd.org/dataoecd/14/31/11728549.pdf.

Heller, M. and R. Eisenberg. (1998). "Can Patents Deter Innovation: The Anticommons in Biomedical Research" *Science* 280:698-701.

Heffan, I. (1997). "Willful Patent Infringement." *Federal Circuit Bar Journal* 7(Summer):115.

Henderson, R., L. Orsenigo, and G. Pisano. (1999). "The Pharmaceutical Industry and the Revolution in Molecular Biology: Interactions among Scientific, Institutional and Organizational Change." In *Sources of Industrial Leadership: Studies of Seven Industries*, D. C. Mowery and R. R. Nelson, eds. New York: Cambridge University Press.

Henry, M., M. Cho, M. Weaver, and J. Merz. (2002). "DNA Patenting and Licensing." *Science* 297(5585):1279.

Hicks, D. (1999). "Innovation in Information Technology in the United States: A Portrait Based on Patent Analysis." CHI Research Working Paper.

Hicks, D., T. Breitzman, D. Olivastro, and K. Hamilton. (2001). "The Changing Composition of Innovative Activity in the US—A Portrait Based on Patent Analysis." *Research Policy* 30(4):681-704.

Hofer, R. and L. Fitzgerald. (1995). "New Rules for Old Problems: Defining the Contours of the Best Mode Requirement in Patent Law." *American University Law Review* 44:2309.

Horstmann, I., G. M. MacDonald, and A. Slivinski. (1985). "Patents as Information Transfer Mechanisms: To Patent or (Maybe) not to Patent." *Journal of Political Economy* 93(5):837-858.

Hunt, R. (1999). "Non-obviousness and the Incentive to Innovate: An Economic Analysis of Intellectual Property Reform." Federal Reserve Bank of Philadelphia Working Paper No. 99-3. Available at http://www.phil.frb.org/files/wps/1999/wp99-3.pdf.

Hunt, R. (2001). "You Can Patent That? Are Patents on Computer Programs and Business Methods Good for the New Economy?" *Philadelphia Federal Reserve Bank Business Review* 2001(Q1):5-15.

Institute of Medicine. (2003). *Large-Scale Biomedical Science: Exploring Strategies for Future Research.* Washington, D.C.: The National Academies Press.

Institute of Medicine. (Forthcoming). *Shortening the Timeline for New Cancer Treatments.* Washington, D.C.: The National Academies Press.

Intellectual Property Owners Association. (2003). "Statement of Intellectual Property Owners Association to the House Judiciary Subcommittee on Courts, the Internet and Intellectual Property (Apr. 3)." Available at http://www.ipo.org/AMTemplate.cfm?Section=IPO_Position_Statements1&Template=/ContentManagement/ContentDisplay.cfm&ContentID=8383.

Jaffe, A. (1999). "The U.S. Patent System in Transition: Policy Innovation and the Innovation Process." NBER Working Paper 7280. Available at http://www.nber.org/papers/w7280.

Jaffe, A. (2000). "The U.S. Patent System in Transition: Policy Innovation and the Innovation Process." *Research Policy* 29:531-557.

Jaffe, A. and J. Lerner. (2004). *Innovation and Its Discontents: How Our Broken Patent System Is Endangering Innovation and Progress, and What to Do About It.* Princeton, N.J.: Princeton University Press.

Janis, M. (2002). "Patent Abolitionism." *Berkeley Technology Law Journal* 17(2):899-952.

Japan Patent Office, European Patent Office, and U.S. Patent and Trademark Office. (2001). "Trilateral Statistical Report." Available at http://www.european-patent-office-org/tws/tsr_2001/index.php.

Jorgenson, D. and K. Stiroh. (2002) "Raising the Speed Limit: U.S. Economic Growth in the Information Age." In *Measuring and Sustaining the New Economy*, D. Jorgenson and C. Wessner, eds. Washington, D.C.: National Academy Press.

Kastriner, L. (1991). "The Revival of Confidence in the Patent System." *Journal of the Patent and Trademark Office Society* 73:5-23.

King, J. (2003). "Patent Examination Procedures and Patent Quality." In *Patents in the Knowledge-Based Economy*, W. Cohen and S. Merrill, eds. Washington, D.C.: The National Academies Press.

Koenig, G. (1980). *Patent Invalidity: A Statistical and Substantive Analysis.* New York: Clark Boardman.

Korn, D and S. J. Heinig, eds. (2002). "Public versus Private Ownership of Scientific Discovery: Legal and Economic Analyses of the Implications of Human Gene Patents." *Academic Medicine: Journal of the Association of American Medical Colleges* 77(Dec.):1301-1308.

Kortum, S. and J. Lerner. (1998). "Stronger Protection or Technological Revolution: What is Behind the Recent Surge in Patenting?" *Carnegie-Rochester Conference Series on Public Policy* 48:247-304.

Lanjouw, J. and I. Cockburn. (2000). "Do Patents Matter? Empirical Evidence After GATT." NBER Working Paper No. 7495. Available at http://ssrn.com/abstract=214635.

Lanjouw, J. and J. Lerner. (1997). "The Enforcement of Intellectual Property Rights: A Survey of the Empirical Literature." NBER Working Paper No. W6296. Available at http://ssrn.com/abstract=226053.

Lanjouw, J. and M. Schankerman. (2003). "Enforcement of Patent Rights in the United States." In *Patents in the Knowledge-Based Economy*, W. Cohen and S. Merrill, eds. Washington, D.C.: The National Academies Press.

Laurie, R. and R. Beyers. (2001). "The Patentability of Internet Business Methods: A Systematic Approach to Evaluating Obviousness." *Journal of Internet Law* 4(11).

Lemley, M. (2001). "Rational Ignorance at the Patent Office." *Northwestern University Law Review* 95(4):1495-1532.

Lemley, M. (2002). "An Empirical Study of the 20-Year Patent Term." In *The Economics of Intellectual Property,* R. Towse and R. W. Holzhauer, eds. Northampton, Mass.: Edward Elgar Publishing.

Lemley, M. and C. Chien. (2003). "Are the U.S. Patent Priority Rules Necessary?" *Boalt Working Papers in Public Law,* no. 32. Available at http://repositories.cdlib.org/boaltwp/32.

Lemley, M. and K. Moore. (2004). "Abolishing Patent Continuations." *Boston University Law Review* 84(Feb.):101.

Lemley, M. and R. Tangri. (2003). "Ending Patent Law's Willfulness Game." *UC Berkeley Public Law and Legal Theory Research Paper Series,* no. 142.

Lerner, J. (1995). "Patenting in the Shadow of Competitors." *Journal of Law and Economics.* 38(Oct.):463-495.

Lerner, J. (2002). "150 Years of Patent Protection." *American Economic Review* 92(2):221-225.

Levin, J. and R. Levin. (2003). "Benefits and Costs of an Opposition Process." In *Patents in the Knowledge-Based Economy*, W. Cohen and S. Merrill, eds. Washington, D.C.: The National Academies Press.

Levin, R. (1982). "The Semiconductor Industry." In *Government and Technical Progress: A Cross-Industry Analysis*, R. R. Nelson, ed. New York: Pergamon Press.

Levin, R., A. Klevorick, R. Nelson, and S. Winter. (1987). "Appropriating the Returns from Industrial R&D." *Brookings Papers on Economic Activity* 3:783-820.

Lunney, G. (2001). "E-Obviousness." *Michigan Telecommunications and Technology Law Review* 7(363):363-421. Available at http://www.mttlr.org/volseven/LunneytypeRE-PDF.pdf.

Mansfield, E. (1986). "Patents and Innovation: An Empirical Study." *Management Science* 32:173-181.

Merges, R. (1994). "Intellectual Property Rights and Bargaining Breakdown: The Case of Blocking Patents." *Tennessee Law Review* 62(1):74-106.

Merges, R. (1999). "As Many as Six Impossible Patents Before Breakfast: Property Rights for Business Concepts and Patent System Reform." *Berkeley Technology Law Journal* 14:578-615.

Merges, R. and J. Duffy. (2002). *Patent Law and Policy: Cases and Materials*. 3rd Edition, New York: Matthew Bender.

Merges, R. and R. Nelson. (1990). "On the Complex Economics of Patent Scope." *Columbia Law Review* 90(4):839-916.

Merton, R. (1973). *The Sociology of Science: Theoretical and Empirical Investigations*. Chicago: University of Chicago Press.

Meurer, M. (1989). "The Settlement of Patent Litigation." *RAND Journal of Economics* 20:77-91.

Merz, J. F, D. G. Kriss, D. D. G. Leonard, and M. K. Cho. (2002). "Diagnostic Testing Fails the Test." *Nature* 415(7 Feb.):577-579.

Moser, P. (2003). "How Do Patent Laws Influence Innovation? Evidence from Nineteenth Century World Fairs." NBER Working Paper No. w9909. Available at http://www.nber.org/papers/w9909.

Mossinghoff, G. (2002). "The First-to-Invent System Has Provided No Advantage to Small Entities." *Journal of the Patent and Trademark Office Society* 84(6):1.

Mossinghoff, G. and V. Kuo. (1998). "World Patent System Circa 20XX, A.D." *IDEA: The Journal of Law and Technology* 38:529-575.

Mossinghoff, G. J. and V. Kuo. (2002). "Post-Grant Review of Patents: Enhancing the Quality of the Fuel of Interest." *IDEA: The Journal of Law and Technology* 43(1):83-110.

Nard, C. (2002). "Toward a Cautious Approach to Obeisance: The Role of Scholarship in Patent Law Jurisprudence." *Houston Law Review* 39(3):667-692.

National Institutes of Health. (1998). "Report of the National Institutes of Health (NIH) Working Group on Research Tools. Presented to the Advisory Committee to the Director, June 4, 1998." Available at http://www.nih.gov/news/researchtools/index.htm.

National Research Council. (1919). *Report of the Patent Committee of the National Research Council*. Reprint and Circular Series, Number 1. Washington, D.C.: National Research Council.

National Research Council. (1936). "Report of the Committee on the Relation of the Patent System to the Stimulation of New Industries." *Journal of the Patent Office Society* 18(2):94-107.

National Research Council. (1997) *Intellectual Property Rights and Research Tools in Molecular Biology*. Washington, D.C.: National Academy Press.

National Research Council. (1999a). *Securing America's Industrial Strength*. Washington, D.C.: National Academy Press.

National Research Council. (1999b). *U.S. Industry in 2000: Studies in Competitive Performance*. Washington, D.C.: National Academy Press.

National Research Council. (2000). *The Digital Dilemma: Intellectual Property in the Information Age*. Washington, D.C.: National Academy Press.

National Research Council. (2001). *Trends in Federal Support of Research and Graduate Education*. Washington, D.C.: National Academy Press.

Nelson, R. (2003). "The Market Economy and the Scientific Commons." Unpublished paper.

Nuffield Council on Bioethics. (2002). *The Ethics of Patenting DNA*. London, UK: Nuffield Council on Bioethics. Available at http://www.nuffieldbioethics.org/publications/pp_0000000014.asp.

O'Donoghue, T., S. Scotchmer, and J.-F. Thisse. (1998). "Patent Breadth, Patent Life, and the Pace of Technological Progress." *Journal of Economic Management and Strategy* 7(1):1-32.

OECD. (2003). "Patents and Innovation: Trends and Policy Changes." DSTI/STP (2003)27. Available at http://www.oecd.org/dataoecd/48/12/24508541.pdf.

O'Rourke, M. (2000). "Toward a Doctrine of Fair Use in Patent Law." *Columbia Law Review* 100(5):1177-1250.

Park, W. and J. Ginarte. (1997). "Intellectual Property Rights and Economic Growth." *Contemporary Economic Policy* 15(Jul.):51-61.

Pollack, A. (2002). "University Resolves Dispute on Stem Cell Patent License." *The New York Times*, January 10, C11.

Pooley. J. (1997-1999). *Trade Secrets*. New York: Law Journal Press.

Poplawski, E. (2001). "Effective Preparation of Patent Related Exculpatory Legal Opinions." *AIPLA Quarterly Journal* 29(3):269-316.

Powers, M. and S. Carlson. (2001). "The Evolution and Impact of the Doctrine of Willful Patent Infringement." *Syracuse Law Review* 51:53-115.

President's Commission on the Patent System. (1966). *"To Promote the Progress of ...Useful Arts" In an Age of Exploding Technology*. Washington, D.C.

Priest, G. (1986). "What Economists Can Tell Lawyers about Intellectual Property." *Research on Law and Economics* 8:19.

Quillen, C. and O. Webster. (2001). "Continuing Patent Applications and Performance of the U.S. Patent Office." *Federal Circuit Bar Journal* 11(1):1-21.

Quillen, C., O. Webster, and R. Eichmann. (2002). "Continuing Patent Applications and Performance of the U.S. Patent and Trademark Office—Extended." *Federal Circuit Bar Journal* 12(1):33-55.

Rai, A. and R. Eisenberg. (2003). "Bayh-Dole Reform and the Progress of Biomedicine." *Law and Contemporary Problems* 66(Winter/Spring):289-314.

Rivette, K. and D. Kline. (2000). *Rembrandts in the Attic: Unlocking the Hidden Value of Patents.* Boston: Harvard Business School Press.

Sakakibara, M. and L. Branstetter. (2001)."Do Stronger Patents Induce More Innovation? Evidence from the 1988 Japanese Patent Law Reforms." *Rand Journal of Economics* 32:77-100.

Saxenian, A. (1996). *Regional Advantage: Culture and Competition in Silicon Valley and Route 128.* Boston: Harvard University Press.

Scherer, F. M., S. Herzstein, Jr., A. Dreyfoos, W. Whitney, O. Bachmann, C. Pesek, C. Scott, T. Kelly, and J. Galvin. (1959). *Patents and the Corporation: A Report on Industrial Technology Under Changing Public Policy.* 2nd ed. Boston: Harvard University, Graduate School of Business Administration.

Scotchmer, S. (1991). "Standing on the Shoulders of Giants: Cumulative Research and the Patent Law." *Journal of Economic Perspectives* 5(1):29-41.

Scotchmer, S. (1996). "Protecting Early Innovators: Should Second Generation Products Be Patentable?" *RAND Journal of Economics* 27(2):322-331.

Shimbo, I., R. Nakajima, S. Yokoyama, and K. Sumikura. (2004). "Patent Protection for Protein Structure Analysis." *Nature Biotechnology* 22(1):109-112.

Stolberg, S. (2001). "Suit Seeks to Expand Access to Stem Cells." *The New York Times*, August 14, C2.

Strandburg, K. (2004). "What Does the Public Get? Experimental Use and the Patent Bargain. " *Wisconsin Law Review* 73(1):81-155.

Taylor, C. and Z. Silbertson. (1973). *The Economic Impact of the Patent System: A Study of the British Experience.* Cambridge, Mass.: Cambridge University Press.

Taylor, E. and G. Von Tersch. (1998). "A Proposal to Shore up the Foundation of Patent Law that the Underwater Line Eroded." *Hastings Communication & Entertainment Law Journal* 20:721.

United Kingdom Royal Society. (2003). "Keeping Science Open: The Effects of Intellectual Property Policy on the Conduct of Science." Available at http://www.royalsoc.ac.uk/files/statfiles/document-221.pdf.

U.S. Court of Claims. (2002). "Rules of the United States Federal Court of Claims." Available at http://www.uscfc.uscourts.gov/rules.htm.

U.S. Department of Commerce, Office of Inspector General. (1990). "Improvements Needed in the Patent Quality Review Program." Audit Report No. EAD-0231-0-0002, Feb.

U.S. Department of Commerce, Office of Inspector General. (1997). "Patent Quality Controls are Inadequate." Audit Report No. PTD-997707-0001, Sept.

U.S. Department of Commerce, Office of Inspector General. (1998). "Board of Patent Appeals and Interferences: High Inventory and Inadequate Monitoring Threaten Effectiveness of Appeals Process." Audit Report No. BTD-10628-8-0001, Sept.

U.S. General Accounting Office. (2001). "State Immunity in Infringement Actions." GAO-01-811, Sept.

U.S. General Accounting Office. (2002). "Federal Action Needed to Help Small Businesses Address Foreign Patent Challenges." GAO-92-789, July.

U.S. Patent and Trademark Office. (1996). "Examination Guidelines for Computer-Related Inventions; Final Version." *Federal Register* 61(40):7478-7492.

U.S. Patent and Trademark Office. (2001). U.S. Patent and Trademark Office Annual Report, Fiscal Year 2000.

U.S. Patent and Trademark Office. (2002a). Manual of Patent Examining Procedure (MPEP). Edition 8 (Aug.). Available at http://www.uspto.gov/web/offices/pac/mpep/index.html.

U.S. Patent and Trademark Office. (2002b). U.S. Patent and Trademark Office Performance and Accountability Report, Fiscal Year 2001.

U.S. Patent and Trademark Office. (2003a). U.S. Patent and Trademark Office Performance and Accountability Report, Fiscal Year 2002.

U.S. Patent and Trademark Office. (2003b). The 21st Century Strategic Plan. Available at http://www.uspto.gov/web/offices/com/strat21/stratplan_03feb2003.pdf.

U.S. Patent and Trademark Office. (2003c). Post-Grant Review of Patent Claims. Available at http://www.uspto.gov/web/offices/com/strat21/action/sr2.htm.

Vermont, S. (2001). "A New Way to Determine Obviousness: Applying the Pioneer Doctrine to 35 U.S.C. § 103(A)." *AIPLA Quarterly Journal* 29(Summer):375-444.

von Hippel, E. (2001). "Innovation by User Communities: Learning from Open Source Software." *Sloan Management Review* 42(4):82-86.

Wagner, R. P. (2003). "Of Patents and Path Dependency: A Comment on Burk and Lemley." *Berkeley Technology Law Journal* 18:1341-1360.

Walsh, J., A. Arora, and W. Cohen. (2003). "Research Tool Patent and Licensing and Biomedical Innovation." In *Patents in the Knowledge-Based Economy*, W. Cohen and S. Merrill, eds. Washington, D.C.: The National Academies Press.

Ziedonis, R. (2003). "Patent Litigation in the U.S. Semiconductor Industry." In *Patents in the Knowledge-Based Economy*, W. Cohen and S. Merrill, eds. Washington, D.C.: The National Academies Press.

Acronyms

ABA	American Bar Association
AIPA	American Inventors' Protection Act of 1999
AIPLA	American Intellectual Property Law Association
APJ	Administrative Patent Judge
BPAI	Board of Patent Appeals and Interferences
CCPA	Court of Customs and Patent Appeals
CMS	Carnegie Mellon Survey
DMCA	Digital Millennium Copyright Act
EPC	European Patent Convention
EPO	European Patent Office
EST	Expressed Tag Sequence
FOA	First Office Action
FTC	Federal Trade Commission
FY	Fiscal Year
GAO	Government Accounting Office
GATT	General Agreement on Tariffs and Trade
IG	Inspector General
IP	Intellectual Property

IPO	Intellectual Property Owners Association
IPR	Intellectual Property Right
IT	Information Technology
JPO	Japanese Patent Office
MIT	Massachusetts Institute of Technology
MOU	Memorandum of Understanding
MPEG	Moving Picture Experts Group
MPEP	Manual of Patent Examining Procedure
NAFTA	North American Free Trade Agreement
NIH	National Institutes of Health
NRC	National Research Council
OECD	Organisation for Economic Co-operation and Development
PCT	Patent Cooperation Treaty
R&D	Research and Development
RCE	Request for Continued Examination
STEP	Board on Science, Technology, and Economic Policy
TRIPS	Trade-Related Aspects of Intellectual Property Rights Agreement
USPTO	U.S. Patent and Trademark Office
WARF	Wisconsin Alumni Research Foundation
WIPO	World Intellectual Property Organization
WTO	World Trade Organization

Law Cases Cited

Anderson's Black Rock, Inc. v. Pavement Salvage Co., 396 U.S. 57, 90 S. Ct. 305, 90 L. Ed. 2d 258, *available at* 1969 U.S. LEXIS 3322, 163 U.S.P.Q. (BNA) 673 (1969).

AT&T Corp. v. Excel Communs., Inc., 172 F.3d 1352, *available at* 1999 U.S. App. LEXIS 7221, 50 U.S.P.Q.2d (BNA) 1447 (Fed. Cir. 1999).

AT&T Corp. v. Excel Communs. Inc., 1999 U.S. Dist. LEXIS 17871, 52 U.S.P.Q.2d 1865 (BNA) (D.Del. Oct. 25, 1999).

Boehringer Ingelheim Vetmedica, Inc. v. Schering-Plough Corp., 68 F. Supp. 2d 508, *available at* 1999 U.S. Dist. LEXIS 16989 (D.N.J. 1999).

Burlington Industries, Inc. v. Dayco Corp., 849 F.2d 1418, *available at* 1988 U.S. App. LEXIS 7938, 7 U.S.P.Q.2d (BNA) 1158 (Fed. Cir. 1988).

College Sav. Bank v. Florida Prepaid Postsecondary Educ. Expense Bd., 527 U.S. 666, 119 S. Ct. 2219, 144 L. Ed. 2d 605, *available at* 1999 U.S. LEXIS 4375, 51 U.S.P.Q.2d (BNA) 1065 (1999).

Cuno Engineering Corp. v. Automatic Devices Corp., 314 U.S. 84 , 62 S. Ct. 37, 86 L. Ed. 58, *available at* 1941 U.S. LEXIS 1250, 1942 Dec. Comm'r Pat. 723, 51 U.S.P.Q. (BNA) 272 (1941).

Dann v. Johnston, 425 U.S. 219, 96 S.Ct. 1393, 47 L. Ed. 2d 692, *available at* 1976 U.S. LEXIS 95, 189 U.S.P.Q. (BNA) 257 (1976).

Dennison Mfg. Co. v. Panduit Corp., 475 U.S. 809, 106 S. Ct. 1578, 89 L. Ed. 2d 817, *available at* 1986 U.S. LEXIS 100, 54 229 U.S.P.Q. (BNA) 478 (1986).

Diamond v. Chakrabarty, 447 U.S. 303, 100 S. Ct. 2204, 65 L. Ed. 2d 144, *available at* 1980 U.S. LEXIS 112, 206 U.S.P.Q. (BNA) 193 (1980).

Diamond v. Diehr, 450 U.S. 175, 101 S. Ct. 1048, 67 L. Ed. 2d 155, *available at* 1981 U.S. LEXIS 73, 209 U.S.P.Q. (BNA) 1 (1981).

Dickerson v. Zurko, 527 U.S. 150 (1999).

Ex parte Erlich, 22 U.S.P.Q.2d 1463 (BNA) (Bd. Pat. App. & Int. 1992).

Ex parte Goldgaber, 1996 Pat. App. LEXIS 4, 41 U.S.P.Q.2d (BNA) 1172 (Bd. Pat. App. & Int. Mar. 25, 1996).

Festo Corp. v. Shoketsu Kinzoku Kogyo Kabushiki Co., 234 F.3d 558, 56 U.S.P.Q.2d 1865 (Fed. Cir. 2000) (en banc), overruled-in-part by 535 U.S. 722, 122 S. Ct. 1831, 152 L.Ed.2d 944 (2002).

Florida Prepaid Postsecondary Educ. Expense Bd. v. College Sav. Bank, 527 U.S. 627, 119 S. Ct. 2199,144 L. Ed. 2d 575, *available at* 1999 U.S. LEXIS 4376, 51 U.S.P.Q.2d (BNA) 1081 (1999).

Gottschalk v. Benson, 409 U.S. 63, 93 S. Ct. 253, 34 L. Ed. 2d 273, *available at* 1972 U.S. LEXIS 129, 175 U.S.P.Q. (BNA) 673 (1972).

Graham v. John Deere Co., 383 U.S. 1, 86 S. Ct. 684, 15 L. Ed. 2d 545, *available at* 1966 U.S. LEXIS 2908, 148 U.S.P.Q. (BNA) 459 (1966).

In re Alappat, 33F.3d 1526, *available at* 1994 U.S. App. LEXIS 21129, 31 U.S.P.Q.2d (BNA) 1545 (Fed. Cir. 1994).

In re Bell, 991 F.2d 781, *available at* 1993 U.S. App. LEXIS 8603, 26 U.S.P.Q.2d 1529 (Fed. Cir. 1993).

In re Deuel, 51 F.3d 1552, *available at* 1995 U.S. App. LEXIS 6200, 34 U.S.P.Q.2d (BNA) 1210 (Fed. Cir. 1995).

In re Dillon, 919 F.2d 688, *available at* 1990 U.S. App. LEXIS 19768, 16 U.S.P.Q.2d (BNA) 1897 (Fed. Cir. 1990).

In re Durden, 763 F.2d 1406, *available at* 1985 U.S. App. LEXIS 15004, 226 U.S.P.Q. (BNA) 359 (Fed. Cir. 1985).

In re Fine, 837 F.2d 1071, *available at* 1988 U.S. App. LEXIS 686, 5 U.S.P.Q.2d (BNA) 1596 (Fed. Cir. 1988).

In re Gay, 309 F.2d 769, *available at* 50 C.C.P.A. 725, 1962 CCPA LEXIS 179, 1962 Dec. Comm'r Pat. 737, 135 U.S.P.Q. (BNA) 311 (C.C.P.A. 1962).

In re Hass, 141 F.2d 127, *available at* 31 C.C.P.A. 903, 1944 CCPA LEXIS 29, 1944 Dec. Comm'r Pat. 242, 60 U.S.P.Q. (BNA) 548 (C.C.P.A. 1944).

In re Henze, 181 F.2d 196, *available at* 37 C.C.P.A. 1009, 1950 CCPA LEXIS 246, 1950 Dec. Comm'r Pat. 319, 85 U.S.P.Q. (BNA) 261 (C.C.P.A. 1950).

In re. Ochiai, 71 F.3d 1565, *available at* 1995 U.S. App. LEXIS 34998, 37 U.S.P.Q.2d (BNA) 1127 (Fed. Cir. 1995).

In re O'Farrell, 853 F.2d 894, *available at* 1988 U.S. App. LEXIS 10951, 7 U.S.P.Q.2d (BNA) 1673 (Fed. Cir. 1988).

In re Papesch, 315 F.2d 381, *available at* 50 C.C.P.A. 1084, 1963 CCPA LEXIS 387, 1963 Dec. Comm'r Pat. 334, 137 U.S.P.Q. (BNA) 43 (C.C.P.A. 1963).

In re Pleuddemann, 910 F.2d 823, *available at* 1990 U.S. App. LEXIS 13089, 15 U.S.P.Q.2d (BNA) 1738 (Fed. Cir. 1990).

In re Zurko, 142 F.3d 1447, *available at* 1998 U.S. App. LEXIS 8811, 46 U.S.P.Q.2d (BNA) 1691 (Fed. Cir. 1998).

Integra LifeSciences I, Ltd. v. Merck KGaA, 331 F.3d 860 (Fed. Cir. 2003).

Jungersen v. Ostby & Barton Co., 335 U.S. 560, 69 S. Ct. 269, 93 L. Ed. 235, *available at* 1949 U.S. LEXIS 3052, 1949 Dec. Comm'r Pat. 520, 80 U.S.P.Q. (BNA) 32 (1949).

Knorr-Bremse Systeme Fuer Nutzfahrzeuge Gmbh v. Dana Corp. 344 F.3d 1336 (Fed. Cir. 2003).

Madey v. Duke University, 307 F.3d 1351, *available at* 2002 U.S. App. LEXIS 20823, 64 U.S.P.Q.2d (BNA) 1737 (Fed. Cir. 2002).

McMullen Assoc. v. State Board of Higher Education, 268 F.Supp. 735, 154 U.S.P.Q. 236 (BNA) (D.C. Or. 1967).

Molins PLC v. Textron, 48 F.3d 1172, *available at* 1995 U.S. App. LEXIS 2959, 33 U.S.P.Q.2d (BNA) 1823 (Fed. Cir. 1995).

Parker v. Flook, 437 U.S. 584, 98 S. Ct. 2522, 57 L. Ed. 2d 451, *available at* 1978 U.S. LEXIS 122, 198 U.S.P.Q. (BNA) 193 (1978).

Regents of the Univ. of Cal. v. Eli Lilly and Co., 119 F.3d 1559, *available at* 1997 U.S. App. LEXIS 18221, 43 U.S.P.Q.2d (BNA) 1398 (Fed. Cir. 1997).

Semiconductor Energy Lab. Co. v. Samsung Elecs. Co., 204 F.3d 1368, *available at* 2000 U.S. App. LEXIS 3164, 54 U.S.P.Q.2d (BNA) 1001 (Fed. Cir. 2000).

State Industries, Inc. v. A.O. Smith Corp., 751 F.2d 1226, *available at* 1985 U.S. App. LEXIS 14682, 224 U.S.P.Q. (BNA) 418 (Fed. Cir. 1985).

State St. Bank & Trust Co. v. Signature Fin. Group, 149 F.3d 1368, *available at* 1998 U.S. App. LEXIS 16869, 47 U.S.P.Q.2d (BNA) 1596 (Fed. Cir. 1998).

United States v. Adams, 383 U.S. 39, 86 S. Ct. 708, 15 L. Ed. 2d 572, *available at* 1966 U.S. LEXIS 2754, 148 U.S.P.Q. (BNA) 479 (1966).

University of Rochester v. G.D. Searle and Co., Inc., 358 F.3d 916, *available at* 2004 U.S. App. LEXIS 2458, 69 U.S.P.Q.2d 1886 (BNA) (Fed. Cir. Feb. 13, 2004).

Walker Process Equipment, Inc. v. Food Machinery & Chemical Corp., 382 U.S. 172, 86 S. Ct. 347, available at 1965 U.S. App. LEXIS 2340, 147 U.S.P.Q. (BNA) 404 (1965).

Westvaco Corp. v. Int'l Paper Co., 991 F.2d 735, *available at* 1993 U.S. App. LEXIS 6712, 26 U.S.P.Q.2d (BNA) 1353 (Fed. Cir. 1993).

WMS Gaming Inc. v. Int'l Game Tech., 184 F.3d 1339, *available at* 1999 U.S. App. LEXIS 16696, 51 U.S.P.Q.2d (BNA) 1385 (Fed. Cir. 1999).

Yarway Corp. v. Eur-Control USA, Inc., 775 F.2d 268, *available at* 1985 U.S. App. LEXIS 15284, 277 U.S.P.Q. (BNA) 352 (Fed. Cir. 1985).

Appendix A

A Patent Primer

Stephen A. Merrill and George C. Elliott[1]

WHAT IS A PATENT?

Intellectual property rights (IPRs) include copyrights, trade secrets, trademarks, and patents. The most common type of patent, and the subject of this primer, is called the **utility patent** to distinguish it from two special classes—**plant** and **design patents**. A utility patent is an exclusive right of limited duration over a new, non-obvious invention capable of practical application. There are four categories of inventions protected by utility patents: **processes**, **machines**, **manufactures**, and **compositions of matter**. The patent law defines "processes" to include new uses for machines, manufacturers, compositions of matter, and materials; and courts have applied dictionary definitions to the meaning of the other statutory categories. However, new technologies seemingly distinct from these definitions have been rationalized into these four classifications. Software, for example, has been patented as either "virtual" machines[2] or processes.[3]

The right—to prevent others from making, using, selling, offering for sale, or importing the patent holder's invention—is granted in return for publication of the invention. A patent contains **claims** setting out the precise legal boundaries of the protection, which applies to anything falling within the scope of the claims, not simply the inventor's exact original work. Having the boundaries defined in

[1]Any opinions expressed in this appendix are solely the authors' and do not necessarily represent the opinions of the U.S. Patent and Trademark Office (USPTO).

[2]*State Street Bank & Trust Co. v. Signature Financial Group*, 149 F.3d 1368 at 1371-72 (Fed. Cir. 1998).

[3]*AT&T Corp. v. Excel Communications, Inc.*, 172 F.3d 1352 at 1361 (Fed. Cir. 1999).

this manner allows others who may make or use similar inventions to know whether they are infringing the patent. They also allow others to **design around** the claims (change and, ideally, improve upon the invention) without infringing the patent.

WHAT IS THE LEGAL AUTHORITY FOR PATENTS?

Based on utilitarian rather than natural-right theory, patents were among the legal concepts introduced to the American colonies by British settlers. Some colonies began to issue patents as early as 1641 and some states continued to do so through the period of the Articles of Confederation. To resolve their growing conflict over patents, the Constitutional Convention of 1789 resolved to create a national system through the Constitution itself, whose Article I, Section 8 authorized Congress to reward exclusive rights for a limited time to authors and inventors "for their respective writings and discoveries." As secretary of state, Thomas Jefferson was responsible for implementing the first Patent Act (1790), although a pro forma registration system was quickly substituted (in 1793) for the original government approval process. Formal examination of applications by professional examiners was introduced by the 1836 revision of the Patent Act. Additional hurdles (for example, what later became known as the "non-obviousness" criterion) were introduced by the courts in the mid- and late-19th century. The contemporary patent system largely reflects the last major revision of the patent statute, the Patent Act of 1952, although some important changes affecting patent term and publication of pre-grant patent applications have been introduced in the last decade.

WHAT MAY BE PATENTED?

The statutory provision on patent-eligible subject matter is brief and has changed little from the version written by Thomas Jefferson: "any new and useful process, machine, manufacture, or composition of matter, or any new and useful improvement thereof."

The effective scope of patenting has been determined primarily by court cases, among them decisions upholding patents on genetically altered living organisms, isolated genes and parts of genes, computer software, and methods of performing business functions. There are no statutory exclusions, although as a result of legislation enacted in 1996 subsequent patents on surgical procedures may not be enforced against individual physicians and as a result of legislation enacted in 1999, accused infringers of business method patents can assert a **prior use defense**. It has been declared USPTO policy not to issue patents on human beings; this was recently codified in statute. And it remains axiomatic that principles, laws of nature, mental processes, intellectual concepts, ideas, natural phenomena, and mathematical formulae are not patentable, although the line between patentable inventions and principles of nature is becoming more difficult to draw.

HOW ARE PATENTS OBTAINED?

Unlike copyrights and trade secrets, which may be asserted by the originator without prior government approval, but like registered trademarks, patents are the product of applications to a government agency (the U.S. Patent and Trademark Office, or USPTO, in the United States), examination by the office, and usually negotiation between the applicant and the examiner over the scope of the claims allowable—a process called **prosecution**. The application is normally prepared by an attorney or by a **patent agent** registered with the USPTO. The **priority** of an application over any other applications for the same invention is established by determining the date of invention (**first-to-invent**), whereas in other countries the date of filing the application establishes priority of one inventor over another who invents the same thing (**first-to-file**).

Upon receipt by the USPTO along with a filing fee, an application is classified by technology and assigned to an examiner in the relevant **art unit** or division of the office. The examiner generally takes up assigned applications in the order received by the office.

The role of the examiner is to

- review the application to determine whether it complies with the basic formal requirements and legal rules.
- determine the scope of the protection claimed by the inventor.
- devise and carry out a search of previously issued patents and other published literature to determine whether the claimed invention is both **novel** and not an obvious extension or variation of what is already known. Patent and nonpatent literature (for example, scientific, technical, business or other published literature) that is relevant to defining the claims or defeating the patent altogether is known as **prior art**. Patent applicants may submit prior art for consideration by the examiner, or the examiner may be aware of pertinent prior art or discover it during the course of the search. Applicants are required to disclose prior art **known to them** that may be material to the examination of their applications. That does not mean that it is incumbent upon applicants to conduct a thorough search for prior art.
- examine the application to determine that the claimed invention was not known and would not have been **obvious to a person of ordinary skill in the art** at the time of invention based on the prior art found during the search, that the invention has **utility**, and that the invention is described in such full, clear, and concise terms as to enable a person of ordinary skill in the art to make and use it. The written description must satisfy a person working in the technology that the inventor had possession of the invention and provide sufficient guidance to enable

BOX A-1
The Statutory Standards for Patentability

Patent law establishes the standards of patentability against which the USPTO measures a patent application. These standards ask whether the claimed invention is

- **patentable subject matter** under 35 U.S.C. § 101 (basically processes, machines, manufactures, and compositions of matter).
- **novel** under 35 U.S.C. § 102, which requires that the invention not be wholly anticipated by prior art or public domain materials.
- **non-obvious** under 35 U.S.C. § 103, which requires the invention to be beyond the ordinary abilities of a skilled artisan knowledgeable in the appropriate field.
- **useful** under 35 U.S.C. § 101, which means the invention must be minimally operable toward some practical purpose.
- whether the application meets the **disclosure requirements** under 35 U.S.C. § 112 by: (i) so completely describing the invention that skilled artisans are enabled to practice it without undue experimentation; (ii) providing a description sufficient to ensure that the inventor actually has invented what the patent application claims; and (iii) containing distinct, definite claims that set out the proprietary interest asserted by the inventor.

that person to carry out the invention without undue further experimentation or invention.

Upon completion of the initial search and examination, whose duration varies among technologies, the examiner issues an official letter, known as a **First Office Action**, either allowing claims or rejecting them as unpatentable under one or more of the patent statutes.

On average, first actions are now occurring approximately 14.4 months after filing,[4] but it can take years or as little as a few months. The action may be to accept all claims in the application as patentable and allow the application. Normally, however, some or all claims are initially rejected and the applicant is

[4]Department of Commerce. USPTO FY2001 Annual Program Performance Report and 2003 Annual Performance Plan, p. 281.

given time to respond. Sometimes, negotiation between the applicant and the examiner ensues by telephone or in a face-to-face interview. That process may lead the applicant to amend some of the claims or the examiner to amend or withdraw the rejections. In most cases, all issues are not resolved through negotiation and the applicant must reply to the First Office Action with a written response that addresses each ground of rejection by amending the claims and/or providing an argument or evidence to show why the rejections no longer apply to the claimed invention.

If, on the basis of the applicant's response or ensuing negotiation, all claims are determined to be patentable, the examiner allows the application and a patent issues. If agreement cannot be reached on some claims, the examiner issues a **Second Office Action** finally rejecting the unpatentable claims. Following a **final** rejection, the applicant may cancel claims, amend them,[5] appeal the rejections, abandon the application, or file a request for continued examination (RCE). An RCE automatically removes the finality of the previous office action so that the examination process may restart while building on the previous prosecution. An RCE counts for bookkeeping purposes as a new application. Except in the case of an RCE, the entire time that the examiner spends on a single application, from initial search and examination to allowance, appeal or abandonment, averages 20 hours,[6] although the time varies among technologies.

If agreement is not reached on allowable claims, an applicant may file a **continuation** of the original application and obtain an additional round of examination and negotiation. At the time of filing the continuation application, the applicant may add new material ("**new matter**") to the specification of the invention and claim the invention with additional elements that were not originally disclosed, in which case the refiled application is called a **continuation-in-part** application. The original filing date of the parent application is preserved as long as no "new matter" is necessary to support the claimed invention. If additional claimed elements do require new matter for description or enablement, those claims only benefit from the filing date of the continuation-in-part application.

If the examiner concludes that the application contains claims to more than one patently distinct invention (not simply variations on a single invention), the examiner may issue a **restriction** requirement, forcing the applicant to decide which claims to pursue in the original application. Excluded claims, along with an appropriate specification, may be filed separately in a **divisional** application, which receives full benefit of the original filing date as its effective filing date. In certain circumstances an applicant may request a rejoinder of excluded claims,

[5]At this point in the prosecution, amendments are not entered as a matter of right but only at the discretion of the examiner if they raise no new issues and either make the claims allowable or simplify the issues for appeal.

[6]Patent Office Professionals' Association Newsletter. (2001). June/July 01 Vol. 1 No. 5. See http://www.popa.org/newsletters/junjul01.shtml.

thereby maintaining the rejoined claims in the first application without filing a divisional application.

A decision to allow some claims combined with cancellation of any claims that are not allowable results in an issued patent, on average 24.7 months from filing.[7] A record of the **prosecution history** of any patent application is kept in the application file, which may be a paper or electronic file. A rejection that is maintained by the examiner after the applicant has responded to it may be appealed, first to the USPTO's internal Board of Patent Appeals and Interferences (BPAI), a panel of **administrative patent judges** and if not successful there, to the Court of Appeals for the Federal Circuit ("Federal Circuit"). Alternatively, a disappointed applicant who has received an adverse BPAI opinion may file a civil suit against the director of the USPTO in the U.S. District Court for the District of Columbia. An applicant's ceasing to prosecute an application at any point for any reason is known as **abandonment**.

This sequence of steps is illustrated in Figure A-1.

HOW LONG ARE PATENTS EFFECTIVE?

The date of original patent application filing starts the clock on a 20-year period in which any allowed claims resulting from the original application are effective. Until recently the period was 17 years from patent issuance. In certain circumstances pharmaceutical product patents and some other patents that issue after extended administrative delay may be extended beyond that term. And since the 1999 American Inventors' Protection Act (AIPA), a patent may be extended if certain administrative delays that are beyond the applicant's control occur in the USPTO. The AIPA attempted to ensure that inventors would not get less than 17 years of patent protection unless they delayed the prosecution of the application.

To take advantage of the available patent life, the patent holder must continue to pay **maintenance fees** to the USPTO at regular intervals—the end of the third year, the end of the seventh year, and the end of the eleventh year. Failure to pay the fee results in the patent's expiring.

WHAT IS THE DISCLOSURE FUNCTION OF PATENTS?

As a condition of the right to exclude, the issued patent and its allowed claims are published, denying the patent holder the ability to keep the invention secret. Patents are part of the public technical literature because of the requirement that a patent be written in sufficiently clear, concise, exact, and complete terms to enable someone of ordinary skill in the art to make and use the invention.

---

[7]Department of Commerce. USPTO FY2001 Annual Program Performance Report and 2003 Annual Performance Plan, p. 282.

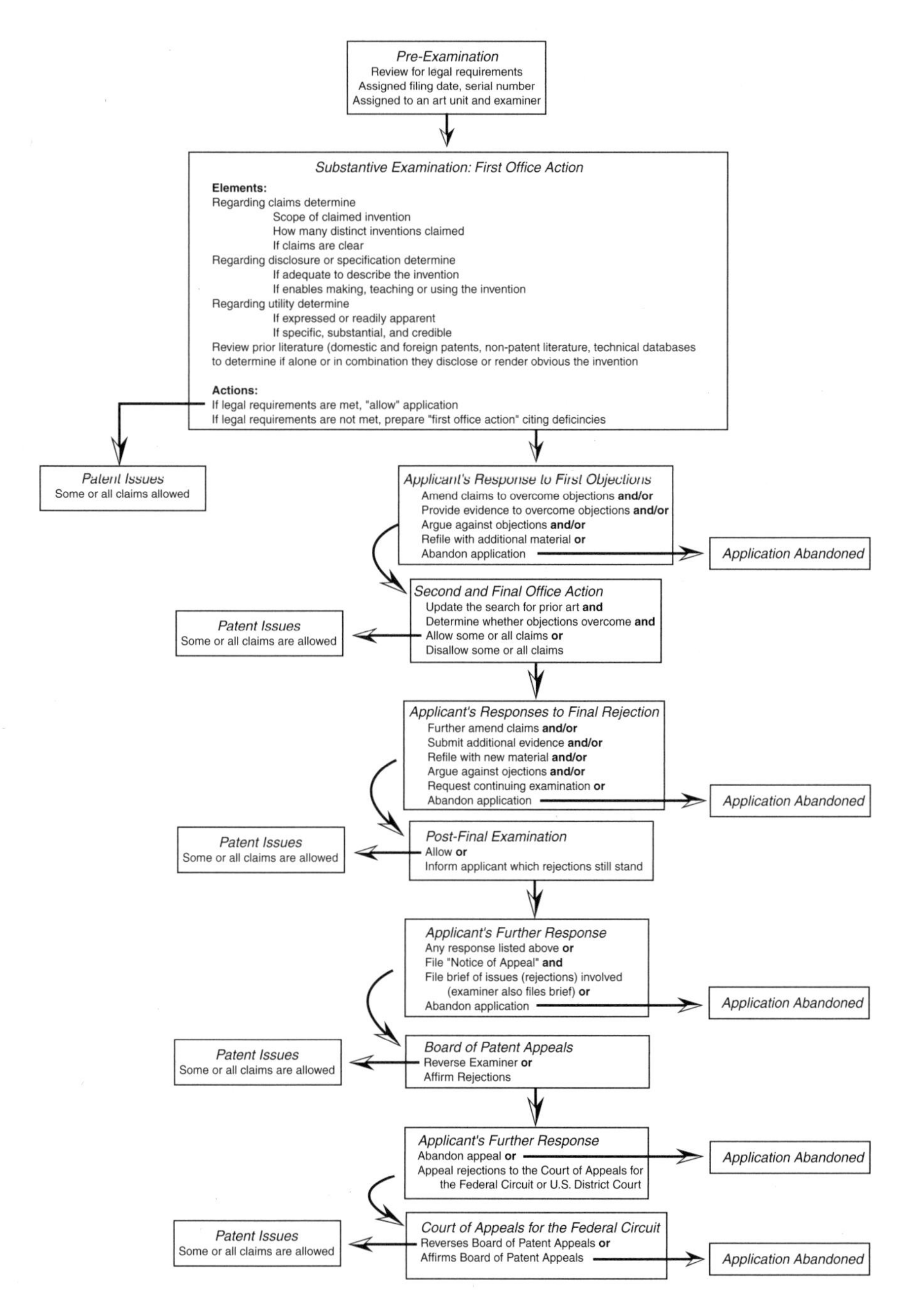

FIGURE A-1 Patent application examination process.

Since March 2001, as the result of an international agreement to harmonize certain administrative practices, the United States has published most but not all pending patent applications after 18 months. Applicants who agree not to file for a patent outside of the United States may withhold their applications from publication until a patent is issued. Published U.S. applications disclose the name of the inventor(s), patent attorney or correspondence address, assignee, the entire specification as filed, and certain materials that may be submitted after filing but are necessary to complete the application; they do not reveal the results of the examiner's prior art search.

WHO OWNS A PATENT?

A patent is issued to the inventor(s) named on the patent application, but **title** is frequently **assigned** to an employer (for example, firm, university, laboratory) or sponsoring organization as part of a prior agreement governing work products. Patents, like other property, may be sold outright, given away, or transferred along with other corporate assets in a merger or acquisition.

HOW ARE PATENTS USED?

There are several ways a patent owner exploits the economic value of a patent or a **portfolio** of related patents. All depend upon the threat of damages or injunction or both that may result from a suit for infringement, but their relative importance or incidence depends greatly upon the technology involved, the market, the scope of the patent claims, and the goals of the patent holder and competitors.

- A patent may be used to deter a potential competitor from entering the market rather than risk losing the investment necessary to do so.
- A patent can make market entry slower, less effective, or more costly by forcing a competitor to evaluate the patent and/or attempt to design around it.
- Many patents are licensed by their owners to other parties for commercial use. A patent may be **licensed exclusively**, giving one party the sole right to use the invention for any purpose, licensed exclusively for a particular **field of use**, or **licensed nonexclusively**, meaning that the owner and other licensees can use the invention. Licensees normally pay patent owners **fees** or **royalties**, although a **cross-licensing** arrangement may give reciprocal rights without payment of money. In that case the value derived is in obtaining access to a needed technology. The right to exclude, however, implies a right to refuse to license a patented invention. With rare exceptions, licensing cannot be compelled in the United States.
- Patents may confer leverage in other negotiations, for example, in setting technical standards for products in an industry requiring compatibility of component systems.

• Patents are among a company's **intangible assets** that may enhance its attraction to investors or its valuation in a merger or acquisition.

• An increasingly common use of patents is defensive. If a patent holder faces the possibility of being sued for infringement, having the ability to make a counter claim for infringement of the patent holder's own patent may help to avoid paying licensing fees or being sued or may induce settlement of an infringement claim.

• Barring settlement, patent suits can result in injunctive relief stopping the accused activity or substantial damage awards if infringement is established.

HOW IS A PATENT CHALLENGED?

A patent application or an issued patent may be challenged administratively (that is, within the USPTO) on narrow grounds. One basis for challenge is that another applicant may have made the invention first. Such a claim results in an **interference proceeding** conducted by the BPAI to establish who was the first inventor. Interferences do not occur in other countries' first-to-file systems where priority is simply a function of the order in which applications are received.

Once a patent is issued, the patentee, a third party, or rarely, the director of the USPTO may seek to have it re-examined if a substantial question of patentability based on prior art is raised. The relevant prior art is no longer limited to prior art that was not uncovered and therefore not considered in the initial examination. **Re-examination** may occur at any time during the life of a patent and is performed by a different examiner than the original one, and its outcome can be appealed to the BPAI. There are two types of re-examination proceedings, one in which there is no third-party participation (***ex parte* re-examination**) and one available for patents applied for after November 29, 1999, in which a third-party complainant can participate (***inter partes* re-examination**). Until November 2002, third-party *inter partes* re-examination requesters were barred from appealing a decision of the Board of Patent Appeals and Interferences. As a result, third-party requests have been very infrequent. The 21st Century Department of Justice Appropriations Authorization Act, Public Law 107-273, gives third-party *inter partes* requesters the ability to appeal board decisions to the Court of Appeals for the Federal Circuit. Nearly one-half of *ex parte* re-examinations are brought by patent owners seeking to strengthen at least a portion of their own rights with or without a narrowing amendment because some prior art has come to light.

In Europe and some other countries it is possible for third parties to challenge patent validity on almost any of the grounds considered in examination, but these patent **oppositions** must be initiated within a few months after the patent is issued. Nevertheless, oppositions are substantially more frequent than U.S. re-examinations.

A patent holder may be sued in federal court to have the patent declared invalid so long as the patentee has made an actual or implicit charge of infringement and the complainant is in a position to engage in an allegedly infringing activity. Accused **infringers** (defendants) almost invariably challenge the validity of the patent at issue in litigation. Absent an infringement allegation, usually first embodied in a cease and desist letter from the patent holder's attorney, there is no cause of action giving a party standing to bring a patent validity suit. Although appeals from USPTO decisions to reject patent applications also result in judicial review of patentability, third parties (potential infringers) may not participate in appeals from regular examinations or *ex parte* re-examination.

HOW IS A PATENT ENFORCED?

The right to exclude is a right to sue to stop the unauthorized making, using, selling, or offering for sale of something within the scope of the claims of a patent or to seek **damages** for the infringement or both **injunctive relief** and damages. A patent can be infringed under either of two doctrines—**literal infringement**, where every claim element is literally present in the accused device, product, or method, or infringement under the **doctrine of equivalents**, where there is an equivalent to the missing element in the accused device, product, or method. There are also three types of infringement: (1) A **direct infringer** is one who is actually using the accused product or practicing the accused process. (2) A **contributory infringer** does not actually practice the claimed invention but, for example, provides a product that can only be used in an infringing manner. (3) **Inducement** occurs when one actively causes another to infringe. Having a patent portfolio combining product, process, and product-by-process claims potentially enables a patent owner a choice of claims to assert against different infringers. There is a six-year statute of limitations on infringement.

Patent suits are tried in **federal district courts.** Cases involving claims to monetary damages are tried by juries unless the right to a jury trial is waived by both sides. Suits for injunctive relief are equity actions. There is no right to a trial by jury in equity, and sometimes patentees avoid jury trials by foregoing money. In addition to the allegation of infringement and the counterallegation of patent invalidity, questions of intent and state of mind frequently arise in patent suits—that is, whether the infringement was **willful infringement** (knowing without a good-faith belief that the patent is invalid and/or not infringed); whether the applicant misled the USPTO in prosecuting the application, for example, by withholding known prior art (**inequitable conduct**), and whether the patent holder disclosed the **best mode** of implementing the invention.

From a plaintiff's perspective a successful suit results in the award of damages once infringement is established and/or an order to cease the accused activity. Where willful infringement is found, the damage award to the plaintiff may be tripled in value. An injunction may be entered at any stage of litigation and is rarely stayed pending an appeal. Most suits are settled before reaching the point of a decision.

Since its creation in 1982 appeals of district court decisions in patent cases are to the Court of Appeals for the Federal Circuit in Washington, D.C. (rather than the regional appellate courts) and ultimately to the Supreme Court. In part because there is no possibility of conflicts in patent law interpretation among regional appellate courts, the Supreme Court has granted certiorari (review) to relatively few appeals of Federal Circuit decisions, although the number has increased in recent years. In addition to patent cases, the Federal Circuit handles a few other specialized areas of law—claims against the U.S. government and some international trade, contract, and energy cases. The Federal Circuit does not have jurisdiction in other areas of intellectual property, such as trademark or copyright law unless they are coupled with a patent issue.

HOW ARE PATENTS TREATED INTERNATIONALLY?

A patent is territorial, that is, valid and effective only in the country whose government issued it. The fact that commerce, on the other hand, is international led to the Paris Convention of 1883, which established the principle of national treatment and patent priority rules among the signatories. The convention came to be administered by the World Intellectual Property Organization (WIPO), a specialized agency of the United Nations, which made slow progress on further international **harmonization**.

Among other functions, WIPO administers the **Patent Cooperation Treaty** (PCT), effective in 1978, designed to streamline the process of seeking patent protection in many countries. Under PCT an applicant may file an international application, designating any number of PCT-member states in which patent protection is desired, and may obtain a prior art search, which is conducted by one of the principal national patent offices designated as an International Searching Authority. For an additional fee the applicant may obtain an International Preliminary Examination Report prepared by the USPTO or the European Patent Office. The treaty specifies deadlines for completing the search and the examination, if any. WIPO forwards the results, which are merely advisory, along with a national application, to whichever countries the applicant originally designated. A member country may accept the search and examination opinions without further inquiry and issue or deny a patent or it may conduct a *de novo* search and examination or conduct a more abbreviated inquiry. Prosecution of a PCT application in no way precludes prosecuting of one or more national applications simultaneously to avoid the significant erosion of patent term that would occur if the international and national processes proceeded sequentially.

In the late 1980s U.S. and European international businesses and governments sought a global strengthening of intellectual property rights, but in the forum of multilateral trade negotiations under the General Agreement on Tariffs and Trade (GATT) rather than WIPO. In 1993 the Uruguay Round resulted in an agreement on Trade-Related Aspects of International Property **(TRIPS)**, among other trade accords and created the World Trade Organization (WTO) to administer them. Disagreements over intellectual property rights between members are subject to the WTO's dispute resolution procedure. The U.S., European, and Japanese patent offices have meanwhile pursued harmonization on a bilateral and trilateral basis, with the result, for example, that search results on identical applications are shared although not yet mutually accepted.

Appendix B

Contributors

The following people assisted the committee's deliberations in a variety of ways—providing financial support, participating in workshops, speaking at conferences, presenting views at open meetings of the committee, conducting and reporting on research, and providing other valuable information. These contributions were indispensable to the committee's work, and we are very grateful for them. Affiliations are at the time of participation in the project.

Greg Aharonian
Internet Patent News Service

John Allison
University of Texas, Austin

Bob Armitage
Eli Lilly

Ashish Arora
Carnegie Mellon University

Kevin Baer
U.S. Patent and Trademark Office

Robert L. Barchi
The University of Pennsylvania

Robert Barr
Cisco Systems

Erwin Basinski
Morrison & Foerster LLP

Eugene Bauer
Stanford University

Kathy Behrens
Robertson Stephens Investment Management

Lee Bendekgey
Incyte Corporation

Jeff Brandt
Walker Digital Corporation

Timothy Bresnahan
U.S. Department of Justice

Charles Caruso
Merck & Co.

Yar Chaikovsky
Zaplet

Adriana Chiocchi
J.H. Marsh & McLennan

Iain Cockburn
Boston University

Ray Conley
Oak Hill Venture Partners

John Danforth
Rambus

Bill Davidow
Mohr Davidow Ventures

Susan DeSanti
Federal Trade Commission

Hon. Q. Todd Dickinson
Howrey Simon Arnold & White, LLP

Raoul Drapeau
Independent Inventor

Boro Dropulic
VIRxSYS Corporation

Hon. Jon W. Dudas
U.S. Patent and Trademark Office

Hon. Timothy Dyk
Court of Appeals for the Federal Circuit

Richard Ehrlickman
IBM

Hon. T. S. Ellis III
Federal District Court for the Eastern District of Virginia

Joe Farrell
University of California, Berkeley

Don Felch
UOP, Inc.

Maryann Feldman
Johns Hopkins University

Jim Finnegan
Lucent Technologies

Alec French
House Judiciary Subcommittee on Courts, the Internet, and Intellectual Property

Nancy Gallini
University of Toronto

Alfonso Gambardella
University of Urbino

Richard Gilbert
University of California, Berkeley

Scott Giles
House Science Committee

Nicholas P. Godici
U.S. Patent and Trademark Office

Hon. Daniel S. Goldin
National Aeronautics and Space Administration

Stuart Graham
University of California, Berkeley

Hilary Greene
Federal Trade Commission

Gary Griswold
3M Innovative Properties, Co.

Lewis Gruber
Arryx, Inc.

Christian Gugerell
European Patent Office

Dietmar Harhoff
University of Munich

Frank Hecker
CollabNet, Inc.

Markus Herzog
Weickmann & Weickmann, Munich

Diana Hicks
CHI Research, Inc.

Jennifer Horney
Center for the Public Domain

Justin Hughes
U.S. Patent and Trademark Office

Robert Hunt
Philadelphia Federal Reserve Bank

Daniel Hunter
U.S. Patent and Trademark Office

Hon. Susan Illston
Federal District Court for the Northern District of California

Daniel K. N. Johnson
Wellesley College

Elke Jordan
National Human Genome Research Institute

Dale Jorgenson
Harvard University

Brian Kahin
Internet Policy Institute

Don Kash
George Mason University

Michael L. Katz
University of California, Berkeley

Julie Katzman
Senate Judiciary Committee

Esther Kepplinger
U.S. Patent and Trademark Office

John King
U.S. Department of Agriculture

Michael Kirk
American Intellectual Property Law Association

Gert Kolle
European Patent Office

Martin Konopken
Autodesk, Inc.

Samuel Kortum
Boston University

Katharine Ku
Stanford University

Stephen G. Kunin
U.S. Patent and Trademark Office

Jeffrey P. Kushan
Powell, Goldstein, Frazier and Murphy

Jean Lanjouw
Yale University

Eric Larson
Pfizer

Ron Laurie
Skadden Arps

Mark Lemley
University of California, Berkeley

Joan Leonard
Howard Hughes Medical Institute

Josh Lerner
Harvard Business School

Julia Liebeskind
University of Southern California

Nancy Linck
Guilford Pharmaceuticals

Bill Long
Business Performance Research Associates, Inc.

Chuck Ludlam
Biotechnology Industry Organization

Kristina Lybecker
University of California, Berkeley

Peter Lyman
University of California, Berkeley

Michael Lynch
Micron

Ron Marchant
United Kingdom Patent Office

Charles Marmor
U.S. Patent and Trademark Office

Evelyn McConathy
Dilworth Paxson, LLP

Hon. Roderick McKelvie
Federal District Court for the District of Delaware

Peter S. Menell
University of California, Berkeley

Robert Merges
University of California, Berkeley

Hon. Paul Michel
Court of Appeals for the Federal Circuit

Steven W. Miller
Procter & Gamble

Mary Ellen Mogee
Mogee & Associates

Kimberly Moore
George Mason University School of Law

Michael Morgan
Wellcome Trust Genome Campus

Michael Morris
Pharmacia

David Mowery
University of California, Berkeley

Ronald Myrick
General Electric Co.

Lita Nelsen
MIT

Hon. Pauline Newman
Court of Appeals for the Federal Circuit

Maria Nuzzolillo
U.S. Patent and Trademark Office

Sue H. Palk
National Aeronautics and Space Administration

R. Hewitt Pate
Department of Justice

Wayne Paugh
Intellectual Property Owners Association

Bob Potter
Department of Justice

Arati Prabhakar
U.S. Ventures Partners

Jonathan Putnam
Charles River Associates

Cecil Quillen
Cornerstone Research

Laurie Racine
Center for the Public Domain

Mark Radcliffe
Gray Cary Ware & Freidenrich, LLP

Hon. Randall R. Rader
Court of Appeals for the Federal Circuit

Terry Rea
American Intellectual Property Law Association

Dai Rees
European Patent Office

Larry Respess
Ligand Pharmaceuticals

Betsi Roach
American Bar Association

Mark Rohrbaugh
National Institutes of Health

James Rose
Altera Corporation

Michael Roth
Monsanto Company

Annalee Saxenian
Univeristy of California, Berkeley

Mark Schankerman
London School of Economics

Manny Schechter
IBM

F. M. Scherer
Harvard University

Petra Schmitz
European Patent Office

Susanne Scotchmer
University of California, Berkeley

Charles Shank
Berkeley Labs

Elizabeth Shaw
U.S. Patent and Trademark Office

James F. Shekleton
South Dakota Board of Regents

Donald Siegel
University of Nottingham

Hon. Fern Smith
Federal Judicial Center and U.S. District Court for the Northern District of California

Neil Smith
Howard, Rice, Nemerovski, Canady, Falk & Rabkin

Deepak Somaya
University of Maryland

Ronald Stern
Patent Office Professionals' Association

Scott Stern
Northwestern University

Robert Greene Sterne
Sterne, Kessler, Goldstein & Fox, P.L.L.C.

Robert Stoll
U.S. Patent and Trademark Office

Sylvie Strobel
European Patent Office

John ("Jay") Thomas
George Washington University School of Law

Marie Thursby
Purdue University

Emerson Tiller
University of Texas, Austin

Albert Tramposch
George Mason University School of Law

Lawrence Trask
University of California, Berkeley

Jack L. Tribble
Merck & Co. Inc.

Paul Uhlir
The National Academies

Charles Van Horn
Finnegan, Henderson, Farabow, Garrett & Duner, LP

Hal Varian
University of California, Berkeley

Allen Wagner
Consultant

John Walsh
University of Illinois, Chicago

Mark Webbink
Red Hat, Inc.

Frank Weiss
Carr & Ferrell

John Wetherell
Fish & Richardson, PC

Brian Wright
University of California, Berkeley

Douglas Wyatt
Wyatt, Gerber & O'Rourke

Bob Young
Red Hat, Inc.

Arvids Ziedonis
University of Pennsylvania

Rosemarie Ziedonis
University of Pennsylvania

Thomas Zindrick
Amgen

Harriet Zuckerman
Andrew W. Mellon Foundation

APPENDIX C

Committee and Staff Biographies

Richard C. Levin, Yale University, Co-chair

Richard Levin has been president of Yale University since October 1993. He has been a member of the Yale economics faculty since 1974, when he received his Ph.D. from Yale. He received his bachelor's degree in history from Stanford University and earned a B. Litt. degree in philosophy and politics from Oxford University. A specialist in the economics of technological change, Dr. Levin has written extensively on the patent system, industrial development, the effects of public policy on private industry, and industrial organization. In the mid-1980s he directed a major effort to gather evidence on the incentives for manufacturing industries' investments in research and development. In the 1970s and 1980s his series of papers on the Interstate Commerce Commission's regulation of railroads had significant influence on the course of railroad deregulation, especially on the standards for evaluating the economic impact of railroad mergers. He was appointed to the National Academies' Board on Science, Technology, and Economic Policy in 1999.

Mark B. Myers, Wharton School, University of Pennsylvania, Co-chair

Mark Myers is visiting executive professor in the Management Department at the Wharton Business School, the University of Pennsylvania. His research interests include identifying emerging markets and technologies to enable growth in new and existing companies with special emphases on technology identification and selection, product development, and technology competences. Dr. Myers has served on the Board on Science, Technology, and Economic Policy since 1994.

Dr. Myers retired from the Xerox Corporation in 2000 after a 36-year career in its research and development organizations. He was the senior vice president in charge of corporate research, advanced development, systems architecture, and corporate engineering from 1992 until his retirement. His responsibilities included the corporate research centers, PARC in Palo Alto, California, Webster Center for Research & Technology near Rochester, New York, Xerox Research Centre of Canada, Mississauga, Ontario, and the Xerox Research Centre of Europe in Cambridge, U.K., and Grenoble, France. During this period he was a member of the senior management committee in charge of setting the strategic direction of the company. Dr. Myers is chairman of the board of trustees of Earlham College and has held visiting faculty positions at the University of Rochester and at Stanford University. He holds a bachelor's degree from Earlham College and a doctorate from Pennsylvania State University.

John H. Barton, Stanford University Law School

John H. Barton is George E. Osborne Professor of Law Emeritus at Stanford University, where he taught law and technology and a variety of international courses. He has concentrated for many years on the intellectual property aspects of biotechnology as well as on the relationships between intellectual property and antitrust law. Professor Barton recently chaired the United Kingdom Commission on Intellectual Property Rights, appointed by the U.K. Secretary of State for International Development to examine the impact of intellectual property rights on developing nations. He has also advised the World Health Organization, World Bank, U.S. Agency for International Development, and Rockefeller Foundation programs in agricultural biotechnology. He has been a chair of the Department of Agriculture's National Genetic Resources Advisory Council and a member of the National Institutes of Health Recombinant DNA Advisory Committee and the National Institutes of Health Working Group on Research Tools.

Robert Blackburn, Chiron Corporation

Robert Blackburn is vice president and chief patent counsel of Chiron Corporation. With over 20 years of experience in both corporate and private practice, he has worked in biotechnology IP since its very early days. In the early 1980s he drafted the patent recently upheld in the CellPro litigation, and he successfully argued the *Bell* case (obviousness standard for new genes) in the Court of Appeals for the Federal Circuit. He has litigated biotechnology patents on four continents. On behalf of the Biotechnology Industry Organization and other industry coalitions he has been involved in legislative and policy matters, including the Biotechnology Process Patent Act, the GATT/TRIPS implementing legislation, the American Inventors Protection Act of 1999, and several amicus briefings of the Federal Circuit and the Supreme Court. *The American Lawyer* has named

Mr. Blackburn one of the top 45 in-house counsel under the age of 45. Mr. Blackburn is also a distinguished scholar at the Berkeley Center for Law and Technology, University of California, Berkeley, School of Law (Boalt Hall), a past chairperson of the Intellectual Property Law Committee of the Biotechnology Industry Organization, and a past board member of the Biotechnology Institute. Prior to joining Chiron, Mr. Blackburn was a partner in the northern California office of Irell & Manella; an associate in its predecessor firm, Ciotti & Murashige; assistant patent counsel at Agrigenetics Research Corporation, Boulder, Colorado; and an associate at the law firm of Banner, Birch, McKie & Beckett in Washington, D.C. He received a J.D. from American University, where he was articles editor of the *Law Review*, and a B.S. in chemistry with honors from Case Western Reserve University.

Wesley Cohen, Duke University

Wesley Cohen is professor of economics and management at the Fuqua School of Business, Duke University, and is a research associate of the National Bureau of Economic Research in Cambridge, Massachusetts. Until September 2002 he was professor of economics and social science in the Department of Social Sciences at Carnegie Mellon University and held faculty appointments in its Department of Engineering and Public Policy and its Heinz School of Policy and Management. Focusing on the economics of technological change, Dr. Cohen's research over the past 15 years has explored the determinants of industrial R&D. He has examined the links between firm size, market structure and innovation, firms' abilities to exploit outside knowledge, the knowledge flows affecting innovation, the means that firms use to protect their intellectual property, and the links between university research and industrial R&D, among other subjects. Recently, he coordinated a major survey research study comparing the nature and determinants of industrial R&D in the United States and Japan. He is currently engaged in a multiyear National-Science-Foundation-funded research project on the impact of patenting on innovation. He received his Ph.D. in economics from Yale in 1981.

Frank Collins, ZymoGenetics

Frank Collins is senior vice president of research at ZymoGenetics. He has over 20 years of experience in drug discovery and development. His accomplishments include discovery of a key target in Alzheimer's disease and of new proteins that regulate the nervous system. Previously Dr. Collins was vice president of neuroscience at Amgen, Inc., and vice president of neuroscience at Synergen, Inc. He developed and oversaw new therapeutic programs at both companies, including a 150-person research team at Amgen working in neurological disorders such as Alzheimer's and Parkinson's disease, pain and stroke, as well as metabolic disorders, including obesity, diabetes, dyslipidemias, and cachexia. His academic background includes positions as director of developmental neurobiology at the National Science Foundation and associate professor of anatomy and neurobiology at the University of Utah School of Medicine. Dr. Collins received his M.A. in immunology from the University of California, Berkeley, and his Ph.D. in developmental biology at the University of California, San Diego.

Rochelle Cooper Dreyfuss, New York University School of Law

After spending several years as a research chemist at Vanderbilt University Medical School, the Albert Einstein Medical School, and the Ciba Geigy Corporation, Rochelle Cooper Dreyfuss entered Columbia University Law School, where she was articles and book review editor of the *Columbia Law Review*. Following her graduation in 1981, she became a law clerk first to Chief Judge Wilfred Feinberg of the United States Court of Appeals for the Second Circuit and later to Chief Justice Warren E. Burger of the Supreme Court. In 1983 Ms. Dreyfuss began teaching at the New York University School of Law. Her research and teaching interests include intellectual property, privacy, the relationship between science and law, and civil procedure. She has authored many articles on these subjects and has coauthored casebooks on civil procedure and intellectual property law. Currently she is Pauline Newman Professor of Law. Previously a consultant to the Federal Trade Commission, the Courts Study Committee, and the Presidential Commission on Catastrophic Nuclear Accidents and a member of the Science and Law and Patent Law committees of the Association of the Bar of the City of New York, Ms. Dreyfuss is currently a member of the American Law Institute and a reporter of its Project on Intellectual Property: Principles Governing Jurisdiction, Choice of Law, and Judgments in Transnational Disputes. Her undergraduate degree is from Wellesley College, and she has an M.S. in chemistry from the University of California, Berkeley.

Bronwyn H. Hall, University of California at Berkeley

Bronwyn H. Hall is professor of economics at the University of California, Berkeley, and founder and owner of TSP International, an econometric software firm. She is also a research associate of the National Bureau of Economic Research and the Institute for Fiscal Studies, London. She received a B.A. in physics from Wellesley College in 1966 and a Ph.D. in economics from Stanford University in 1988. Dr. Hall has published numerous articles on the economics and econometrics of technical change. Her current research includes the use of patent citation data for the valuation of intangible (knowledge) assets, comparative firm-level investment studies, measuring the returns to R&D and innovation at the firm level, analysis of technology policies such as R&D subsidies and tax incentives, and studies of the strategic use of patenting in several industries. Dr. Hall was appointed to the National Academies' Board on Science, Technology, and Economic Policy in 1999. Previously she served on the Census Advisory Committee of the American Economic Association. She is currently a member of the International Advisory Board of the New Economic School, Moscow, an associate editor of *Economics of Innovation and New Technology* and the *Journal of Economic Behavior and Organization*, and a member of the editorial board of *International Finance*. She has been a visiting professor at Oxford University and a Hoover Institution national fellow.

Hon. Eugene Lynch, U.S. District Court for the Northern District of California (ret.)

Eugene Lynch is a mediator and arbitrator with JAMS/Endispute, San Francisco, California, where he handles a large number of intellectual property disputes. In addition, he is a member of the Center for Public Resources' National Panel of Distinguished Neutrals. In 1997-1998 he chaired the Kaiser Permanente Blue Ribbon Committee to Reform its Arbitration Procedure. Judge Lynch's judicial career began with his appointment to the San Francisco Municipal Court Bench in 1971. Three years later he joined the San Francisco Superior Court Bench. From 1982 to 1997 he was a judge in the U.S. District Court for the Northern District of California, one of the principal venues of intellectual property litigation in the federal court system. Judge Lynch is a graduate of Santa Clara University and the University of California Hastings College of the Law.

Daniel P. McCurdy, ThinkFire, Ltd.

Dan McCurdy is president and chief executive officer of ThinkFire, Ltd., an adviser on intellectual property matters to firms primarily in the information technology and communications industries. Previously Mr. McCurdy was president of Lucent Technologies' Intellectual Property Business, responsible for protecting, managing, and extracting value from Lucent's intellectual property assets worldwide. Before joining Lucent, he was vice president, Life Sciences, at IBM, where he directed the company's strategy, product and business development, and marketing related to the life sciences industry. In the late 1990s Mr. McCurdy was vice president for corporate development at CIENA Corporation, a publicly traded telecommunications firm. As a member of CIENA's senior management team, he was responsible for mergers, acquisitions, strategic investments, licensing, and corporate partnerships. From 1983 to 1997 Mr. McCurdy served in various positions with IBM. In his last position as director of business development and market strategy for IBM Research he was a member of the 14-person executive team guiding the division. There he was responsible for all intellectual property licensing activities as well as the creation of a variety of joint ventures and technology-based spin-offs. He is a 1981 graduate of the University of North Carolina, Chapel Hill.

Hon. Gerald J. Mossinghoff, Oblon, Spivak, McClelland, Maier & Neustadt (Committee member until December 2003)

Gerald Mossinghoff, a former assistant secretary of commerce and commissioner of patents and trademarks and a former president of the Pharmaceutical Research and Manufacturers of America, is senior counsel to the firm of Oblon, Spivak, McClelland, Maier & Neustadt, where he advises on a broad range of intellectual property matters, including international, legislative, and policy issues. He has been an expert witness in dozens of patent cases in the federal courts. He is also Ciefelli Professor of Intellectual Property Law at the George Washington University Law School, a Distinguished Adjunct Professor at the George Mason University School of Law, and a fellow of the National Academy of Public Administration. At the USPTO Mr. Mossinghoff advised President Reagan on the establishment of the Court of Appeals for the Federal Circuit and initiated a far-reaching automation program of the office's databases. He has served as U.S ambassador to the Diplomatic Conference on the Revision of the Paris Convention and as chairman of the General Assembly of the United Nations World Intellectual Property Organization. Previously he was deputy general counsel of the National Aeronautics and Space Administration. Mr. Mossinghoff received his J.D. with honors from the George Washington School of Law and an electrical engineering degree from St. Louis University.

Gail K. Naughton, Dean, San Diego State University College of Business Administration

Gail Naughton assumed her present position in June 2002. Previously she was vice chairman of the board of directors, and she was a director of La Jolla-based Advanced Tissue Sciences and a director since the firm's inception in 1987. She cofounded the company and was instrumental in taking it public. As the scientific founder and later in various executive positions including president, she set the overall scientific direction for the company while playing a key role in building the company and its management team, raising capital, and increasing public awareness of the company as a pioneer in developing innovative products for patients needing replacement tissues and organs. Dr. Naughton has published extensively in the field of tissue engineering and holds more than 70 U.S. and foreign patents. In 2000 she was the first woman individually to win the National Inventor of the Year Award of the Intellectual Property Owners Association. Dr. Naughton received her bachelor's degree in biology from St. Francis College in New York in 1976, her master's degree in histology (the study of human tissue structure) in 1978, and her Ph.D. in basic medical sciences from New York University in 1981. She completed her postdoctoral training at the New York University Medical Center in the department of dermatology. She served as an assistant professor of research at NYU Medical Center from 1983 to 1985 and as an assistant professor of biology at the City University of New York's Queensborough Community College from 1985 to 1987. She earned her executive M.B.A. from the University of California, Los Angeles, in 2001.

Richard R. Nelson, Columbia University

Richard Nelson is George Blumenthal Professor of International and Public Affairs, Business and Law at Columbia University's School of International and Public Affairs. He joined the faculty in 1986 after a long tenure as professor of economics at Yale University. Dr. Nelson studies the process of long-range economic change, with particular emphasis on technological advance, evolution of economic institutions, roles of government in a mixed economy, and theories of the firm. He was a principal investigator on both the Yale and the Carnegie Mellon surveys of corporate R&D managers with regard to the use of patents and other methods of appropriating returns to R&D investments in a variety of industries. Dr. Nelson's major publications include *An Evolutionary Theory of Economic Change,* with S.G. Winter; *Government and Technical Progress: A Cross-Industry Analysis; High Technology Policies: A Five Nation Comparison*; and *National Innovation Systems: A Comparative Study.* He received his B.A. from Oberlin College and Ph.D. from Yale University.

James Pooley, Milbank, Tweed, Hadley & McCloy, LLP

Jim Pooley is a senior partner in the Palo Alto office of Milbank, Tweed, Hadley and McCloy, LLP. Mr. Pooley has practiced as a trial lawyer in Silicon Valley for over 30 years, focusing on technology litigation and counseling, and handling hundreds of trade secret and patent matters. He was lead trial counsel for Adobe Systems in its successful defense of software patent claims, recognized by the *National Law Journal* as one of the country's 15 "Top Defense Verdicts" of 1997. Mr. Pooley is a frequent lecturer and prolific writer on the law of trade secrets and patents. He is currently an adjunct professor of law at Boalt Hall School of Law, University of California, Berkeley, a former director of the American Intellectual Property Law Association, and a director and officer of the National Inventors Hall of Fame Foundation. Mr. Pooley is an honors graduate of Lafayette College and of Columbia University School of Law.

William J. Raduchel

William Raduchel was until recently executive vice president and chief technology officer of AOL Time Warner, Inc. He assumed that position in 2001 after performing a similar role at America Online, Inc. He joined AOL in 1999 from Sun Microsystems, Inc., where he was chief strategy officer and a member of the executive committee. In his 11 years at Sun he was also chief information officer, chief financial officer, acting vice president of human resources, and vice president of corporate planning and development. Prior to his tenure at Sun, Dr. Raduchel held senior executive positions at Xerox Corporation and McGraw-Hill, Inc. He received his undergraduate degree from Michigan State University and A.M. and Ph.D. degrees in economics from Harvard. He was named to the National Academies' Board on Science, Technology, and Economic Policy in 2000 and currently serves on another National Academies' panel on Internet navigation and domain names.

Pamela Samuelson, University of California, Berkeley, Law School

Pamela Samuelson is Chancellor's Professor of Law and Information Management at the University of California, Berkeley (Boalt Hall), a director of the Berkeley Center for Law and Technology, and an honorary professor at the University of Amsterdam. She came to Boalt in 1996 from the University of Pittsburgh School of Law, where she had taught since 1981. She has also practiced with the New York firm of Willkie Farr and Gallagher and served as a principal investigator for the Software Licensing Project at Carnegie Mellon University. Professor Samuelson has lectured widely and published extensively in the areas of copyright law and software protection. In 1997 she was named a fellow of the John D. & Catherine T. MacArthur Foundation, and in 2000 she was named as one of the 100 most influential lawyers in the United States by the *National Law Journal.* She was elected to membership in the American Law Institute and named a fellow of the Association of Computing Machinery. She has been a contributing editor to the *Communications of the ACM* since 1990. Professor Samuelson received her B.A. and M.A. from the University of Hawaii and her J.D. from Yale Law School.

STAFF

Stephen A. Merrill, Project Director

Stephen Merrill has been executive director of the National Academies' Board on Science, Technology, and Economic Policy (STEP) since its formation in 1991 and has directed several STEP projects on human resources, tax, and research and development as well as intellectual property policies. He joined the National Academies staff in 1987 as the institution's first director of government affairs and congressional liaison. Previously he was a fellow in international business at the Center for Strategic Studies, where he specialized in technology trade issues. For several years until 1981 Dr. Merrill served on various congressional staffs, most recently that of the Senate Commerce, Science, and Transportation Committee, where he organized the first congressional hearings on international competition in biotechnology and microelectronics and was responsible for legislation on industrial innovation and government patent policy. He holds degrees in political science from Columbia (B.A.), Oxford (M.Phil.), and Yale (M.A. and Ph.D.) Universities.

Craig Schultz, Research Associate

Craig Schultz has been with the National Academies' Board on Science, Technology, and Economic Policy since 1998. He has worked on several STEP projects on human resources, government-industry partnerships, research and development, and intellectual property rights. Prior to joining STEP, Mr. Schultz worked in the Office of the Vice President for Development at the University of Virginia. He holds a B.A., High Honors, from the University of Michigan and an M.A. from the University of Virginia.

Camille Collett, Program Associate (Until September 2002)

Camille Collett is currently program associate with the National Academies' Committee on Science, Engineering, and Public Policy, and was a program associate with the Board on Science, Technology, and Economic Policy until September 2002. Prior to joining the National Academies, Ms. Collett was the Web editor for the launch of an alternative health site for women. She has also worked in journal publishing at *The Sciences* and *The Journal of NIH Research.* Ms. Collett is a graduate of the honors English program at the University of Alberta in Edmonton, Alberta, and is currently enrolled in Catholic University Law School.

George Elliott, Department of Commerce Science and Technology Fellow (September 2000 through September 2001)

George Elliott recently assumed the duties of acting director of the Office of Patent Quality Assurance at the United States Patent and Trademark Office, where he is responsible for overseeing a review process aimed at detecting quality problems in allowed applications and applications that are still undergoing examination, and for assisting the Technology Centers in their efforts to improve patent examination quality. He joined the biotechnology examining group at the USPTO as an examiner in 1989 and became a supervisor in 1996, in charge of two art units responsible for applications dealing with gene expression, gene regulation, and antisense therapeutics. As a Department of Commerce science and technology fellow for 2000-2001, Dr. Elliott worked full-time with the staff of the National Academies' Board on Science, Technology, and Economic Policy and its Committee on Intellectual Property Rights in the Knowledge-Based Economy. Opinions expressed by Dr. Elliott during the course of the study and the preparation of this report were his own and not necessarily those of the USPTO. Dr. Elliott received his B.A. in biology from the University of California, San Diego, and his Ph.D. in biology from the University of Utah. Prior to joining the USPTO, he did postdoctoral research at Cambridge University and the University of California, Berkeley.

Russell Moy, Senior Program Officer (From June 2002)

Russell Moy is a senior staff officer in the Board on Science, Technology, and Economic Policy at the National Academies, where he works on issues related to international trade, intellectual property policies, intellectual property enforcement technologies, and technology management. From 2000-2001 Dr. Moy was a policy analyst in the White House Office of Science and Technology Policy, where he supported interagency technology development activities on international trade, health care, and nanotechnology. Earlier Dr. Moy served as a policy analyst in Technology Administration of the U.S. Department of Commerce on the Partnership for a New Generation of Vehicles. Before coming to Washington, D.C., Dr. Moy was the group leader for energy storage programs at Ford Motor Company in Dearborn, Michigan. Dr. Moy holds a J.D. degree from Wayne State University School of Law. He earned Ph.D. and M.S. degrees in chemical engineering from the University of Michigan and a B.S. degree in chemical engineering from Case Western Reserve University.

Aaron Levine, National Research Council Intern (Summer 2003)

Aaron Levine participated in the Christine Mirzayan Internship Program of the National Academies in the summer of 2003. A graduate of the University of North Carolina at Chapel Hill (B.S.) and Cambridge University (M.Phil. in biological sciences), he is currently a Ph.D. student in public policy at Princeton University.

Peter Kozel, National Research Council Intern (Winter-Spring 2004)

Peter Kozel participated in the Christine Mirzayan Internship Program of the National Academies in the winter and spring of 2004. A graduate of the University of Massachusetts (B.S.) and the University of Cincinnati (Ph.D. in molecular genetics), he was an Intramural Research Training Award fellow at the National Institute on Deafness and Other Communication Disorders before coming to the National Research Council.